P.K. Kuroda

The Origin of the Chemical Elements

and the Oklo Phenomenon

with 24 Figures and 48 Tables

Springer-Verlag
Berlin Heidelberg New York 1982

Prof. Paul K. Kuroda,
Dept. of Chemistry, Chemistry Building,
University of Arkansas
Fayetteville, AR 72701, USA

ISBN-13:978-3-642-68669-6   e-ISBN-13:978-3-642-68667-2
DOI: 10.1007/978-3-642-68667-2

Library of Congress Cataloging in Publication Data.
Main entry under title:
Kuroda, P.K. (Paul K.), 1917–
The origin of the chemical elements and the Oklo phenomenon.
Bibliography: p.
Includes index.
1. Chemical elements. 2. Cosmochemistry. 3. Nuclear reactions.
I. Title.
QD 466.K87 1982 564 82-5857
ISBN-13:978-3-642-68669-6 (U.S.) AACR2

© Springer-Verlag Berlin Heidelberg 1982
Softcover reprint of the hardcover 1st edition 1982

Composition: Schreibsatz-Service Weihrauch, Würzburg

2152/3321-543210

To Professor Kenjiro Kimura

Contents

Preface   XI

1. Introduction   1

2. Abundance of the Elements   5

   2.1.   Mendeléeff and the Periodic Law   5
   2.2.   The Ideas of Crookes   6
   2.3.   Richards and Atomic Weights   7
   2.4.   Clarke's Numbers   8
   2.5.   The Rule of Harkins   8
   2.6.   The 1930 Estimates by Noddack   10
   2.7.   The 1938 Estimates by Goldschmidt   12
   2.8.   Geochemical Classification of the Elements   12
   2.9.   The 1956 Estimates by Suess and Urey   12

3. Elements 43 and 61 in Nature   15

   3.1.   The All-Present Theory of Noddack   15
   3.2.   Discoveries of Elements 43 and 61 by Artificial Means   16
   3.3.   Magic Numbers   17
   3.4.   Technetium in Stars   19
   3.5.   Long-Lived Isotopes of Technetium   19
   3.6.   Reported Discoveries of Technetium in Terrestrial Minerals   20
   3.7.   Absence of Primordial Technetium in the Earth's Crust   22
   3.8.   Molybdenum-99 in Non-Irradiated Uranium Salts   26
   3.9.   Technetium in Pitchblende   27
   3.10. Promethium-147 in Non-Irradiated Uranium Salts   28
   3.11. Promethium in Pitchblende   29

4. The Oklo Phenomenon   31

   4.1.   Discovery of Spontaneous Fission   31
   4.2.   Plutonium-239 in Nature   32
   4.3.   Large-Scale Nuclear Processes on the Earth   33
   4.4.   Xenon Isotopes in Radioactive Minerals   34

4.5.	Radioactive Strontium Isotopes in Pitchblende	35
4.6.	Iodine-129 in Pitchblende	37
4.7.	Radioactive Iodine Isotopes in Aqueous Uranium Solutions	37
4.8.	Resonance Capture of Neutrons in Pitchblende	39
4.9.	The Theory of Natural Reactors	40
4.10.	The Uranium-238 to -235 Ratio in Nature	44
4.11.	The Uranium-234 to -238 Ratio in Nature	47
4.12.	Discovery of the Oklo Reactor	48
4.13.	Promethium-147 in the Oklo Reactor	49
4.14.	Plutonium-239 in the Oklo Reactor	50
4.15.	Technetium-99 in the Oklo Reactor	53
4.16.	Search for Additional Natural Reactors	54

5. Synthesis of the Elements in Stars 57

5.1.	Discovery of Helium in the Sun	57
5.2.	The Concept of Frozen Thermodynamic Equilibria	58
5.3.	Deficient Elements	59
5.4.	The Rate of Thermonuclear Reactions	60
5.5.	The C-N Cycle and the Proton-Proton Chain	62
5.6.	Synthesis of the Elements in a Neutron-Rich Environment	63
5.7.	The Big-Bang Theory of Gamow	64
5.8.	The Polyneutron Hypothesis of Mayer and Teller	65
5.9.	The Proton-Neutron Ratio Prior to the Big-Bang	66
5.10.	Theories on the Evolution of Stars	67
5.11.	Supernovae and Californium-254	68
5.12.	Synthesis of the Elements in Stars	70
5.13.	The e-Process According to B_2FH	72
5.14.	The s- and r-Processes According to B_2FH	74
5.15.	Cosmic Black-Body Radiation	78
5.16.	Pulsars or Neutron Stars	79
5.17.	The World of Antimatter	80
5.18.	Possible Climatic Effect of Supernova Explosion	81
5.19.	Search for Neutrinos from the Sun	81
5.20.	Temperature of the Sun	83
5.21.	Further Studies on Nucleosynthesis in Stars	83

6. Plutonium-244 in the Early Solar System 85

6.1.	Rutherford and Soddy's View on the Transuranium Elements	85
6.2.	Rutherford's Calculation of the Age of the Elements	86
6.3.	The Concept of Extinct Radioactivity	87
6.4.	Half-life of Iodine-129 and the Age of the Elements	88
6.5.	Excess ^{129}Xe in Meteorites	89
6.6.	The Plutonium-244 Hypothesis	93
6.7.	Chronology of Nucleosynthesis	100
6.8.	Plutonium-244 in the Early Solar System	102
6.9.	Unsolved Problems in Xenology	107

6.10. Search for Superheavy Elements in Nature 112
6.11. Superheavy Elementary Particles and Quarks in Nature 114

7. Isotopic Anomalies in the Early Solar System 115

7.1.   The Origin of Lithium, Beryllium and Boron 115
7.2.   Isotopic Anomalies in Meteorites 117
7.3.   A Unified Theory of Isotopic Anomalies 120
7.4.   Neon 129
7.5.   Argon 130
7.6.   Krypton 130
7.7.   Xenon 133
7.8.   Barium 140
7.9.   Gadolinium 142
7.10. Other Elements 144

Appendices 147

Appendix I.
        Goldschmidt's table of the abundance of the elements (Originally
        compiled in 1938 and up-dated to 1954) 148

Appendix II.
        The 1956 Suess-Urey abundance table for the individual nuclei 151

Appendix III.
        The 1965 abundance table compiled by Virginia Trimble 158

Subject Index 161

Preface

At about the time I was a student in the 1930's, it had become increasingly evident that all the elements existing on the Earth today had already been discovered. Scientists then began "discovering" new elements by means of their artificial synthesis and some of the man-made elements found important military as well as industrial applications. I have often wondered, however, if the importance of these artificial elements may not have been overly emphasized by contemporary scientists for their practical applications. It seemed to me that these man-made elements were destined to play an important role during the second half of the 20th century in the study of the origin of the elements in the Universe. This subject of study, which dates back to the days of ancient Greek philosophers, may be regarded as the most fundamental in the entire compass of our modern science.

Since I joined the faculty of the University of Arkansas in the early 1950's, I have had the good fortune of being able to maintain a long-range research project, the ultimate goal of which was to elucidate the origin of the elements. I have presented the results from these and related investigations on numerous occasions. While serving as a tour speaker of the American Chemical Society for many years, I have had the privilege of visiting many of the local sections to present a lecture on the origin of the elements. A Japanese-language version of the talk has been presented at universities and research laboratories throughout Japan.

When I was invited to Nagoya University in Japan in the summer of 1977, I was asked by Professor Hideo Yamatera to write a monograph to be published by the Springer Verlag in Germany. The year 1977 happened to be the 100th anniversary of the births of Francis Aston and Frederick Soddy and it appeared to be an appropriate time to undertake such a project.

This monograph could not have been written if it were not for the fact that Professor Yamatera was kind enough to provide strong support for the project. It is a real pleasure to express my deep gratitude to him for his encouragement and patience; and to Dr. F. Boschke of the Springer Verlag for his kindness in deciding to publish the manuscript. Special thanks are also due to the U. S. Atomic Energy Commission and the National Science Foundation for the generous support of the writer's research projects at the University of Arkansas; to Dr. Michiaki Furukawa of Nagoya University for reviewing the manuscript; to Dr. James N. Beck of McNeese State University for his assistance; and to Mrs. Marge Bright for her excellent typing work.

Fayetteville, Arkansas   P.K. Kuroda
May 5, 1981

1.
Introduction

*"It seems not absurd to conceive that at the first produc-
tion of mixed bodies, the universal matter whereof they
among other parts of the universe consisted, was actually
divided into little particles of several sizes and shapes
variously moved."*

Robert Boyle, The Sceptical Chymist, A.D. 1661

One day in June 1936, a notice was placed on the bulletin board of the Chemistry Building of the Imperial University of Tokyo in Japan. It was the announcement that the Nobel
Laureate Francis W. Aston from Cambridge University in England would be presenting a
special lecture on the subject of isotopes at 4:00 p.m. on June 13, 1936. I was a 19-year-
old student then. In order to make sure that I would be able to understand at least a part
of Dr. Aston's lecture, I went to a bookstore in downtown Tokyo and purchased a copy
of "Mass-Spectra and Isotopes" [1.1]. The price of this book was ¥ 12.60 or an équivalent
of $ 6.30. So I told my father that the book was extremely important for my future career
as a scientist and let him pay for it. It turned out that I did not exactly tell him a lie, because I still keep this book in my study and I have read it many times during the past 45
years.

In the Preface of "Mass-Spectra and Isotopes", Aston stated: "I have attempted to
give an account of each individual element relative to its isotopic constitution, a nuclear
'chemistry' which so far as I am aware is the first of its kind. The rapid development of
artificial transmutation of elements appears likely to create a demand for collected data
of this kind in an easily accessible form". I did not realize it then, but I was thus introduced to the field of Nuclear Chemistry when I was only 19. Nuclear Chemistry was soon
destined to have a profound influence on the ultimate fate of my native country Japan, as
well as on my future career as a scientist.

It appeared that Aston was fearful of the possibility that an enormous amount of
energy released by the conversion of matter might somehow engulf the entire earth. In

---

1.1   F.W. Aston, Mass-Spectra and Isotopes, Edward Arnold & Co., London, 1933.

his December 12, 1922, Nobel Lecture, he remarked: "Should the research workers of the future discover some means of releasing this energy in a form which could be employed, the human race will have at its command powers beyond the dreams of scientific fiction: but the remote possibility must always be considered that the energy once liberated will be completely uncontrollable and by its intense violence detonate all neighboring substances. In this event the whole of the hydrogen on the earth might be transformed at once and the success of the experiment published at large to the Universe as a new star".

The fear expressed by Aston may have had a profound effect on the minds of many Japanese scientists in regard to their nuclear weapons program during the war. The fact that many years later I became keenly interested in the study of nuclear chain reactions in nature is no doubt due to my encounter with Dr. Aston in 1936.

In April 1937 I had the good fortune of attending a series of lectures given by Dr. Niels Bohr on the theory of spectra and atomic constitution.

At the end of 1938, Otto Hahn and F. Strassmann [1.2] in Germany discovered fission. In 1939, S. Flügge [1.3] in Germany pointed out the possibility of a self-sustaining uranium chain reaction, which would release enormous amounts of energy. Such a device might be used in future wars. It was noteworthy, however, that he went on even further and suggested the possibility that a self-sustaining uranium chain reaction might have already taken place under natural conditions sometime in the past, possibly in large pitchblende ore deposits at St. Joachimsthal in Bohemia or in carnotite deposits in the Colorado region of the United States.

The 1938 Nobel Prize in Physics was awarded to Enrico Fermi in Italy for his demonstration of the existence of new radioactive elements produced by neutron irradiation, and for his related discovery of nuclear reactions brought about by slow neutrons.

Three years later, Fermi was leading a team of American scientists at the University of Chicago to build the first nuclear reactor. *On December 2, 1942, man achieved here the first self-sustaining chain reaction and thereby initiated the controlled release of nuclear energy* — so reads the plaque at the football stadium of the University of Chicago. Professor Fermi died in Chicago on November 29, 1954. It was about this time when I was beginning to study Fermi's pile theory at the University of Arkansas in an attempt to prove that natural (Pre-Fermi) reactors could have existed on the earth billions of years ago.

During WW II, I was engaged in a geochemical investigation of volcanoes and hot springs — a thesis problem given to me by Professor Kenjiro Kimura of the Imperial University of Tokyo. Thus I became interested in the problems related to large-scale processes occurring in the interior of the earth's crust. I often wondered if the events taking place in the earth's crust might not be *nuclear* rather than *chemical* processes. I also imagined that some hitherto unknown heavy elements which existed in the early history of the earth might have played a role as a source of energy for the volcanic activities.

The events which produced nuclear transformation of an explosive character occurred in the summer of 1945 in Hiroshima and Nagasaki and the war ended abruptly. One day in August 1945, while standing in the ruins of Hiroshima, I became overwhelmed by the power of nuclear energy. The sight before my eyes was just like the end of the world, but I also felt that the beginning of the world may have been just like this. Would it not be

1.2   O. Hahn and F. Strassman, Naturwissenschaften 27:11 (1939).
1.3   S. Flügge, Naturwissenschaften 27:401 (1939).

possible that nuclear chain reactions occurred on the earth then? Would it be absurd to speculate that an exotic element such as plutonium existed on the earth shortly after its birth?

In September 1951, a joint session of the International Union of Pure and Applied Chemistry (IUPAC) and the American Chemical Society (ACS) was held in New York City.

When I had finished with my talk on radium, radon, and their daughters in hot springs in Japan the late Professor Raymond R. Edwards, who was then Chairman of the Chemistry Department of the University of Arkansas offered me a position, and I decided to join the faculty of the University of Arkansas and moved to Fayetteville on June 1, 1952. The study on the radioactivities of the waters of Hot Springs, Arkansas, which began at the University of Arkansas eventually led to the investigations on the occurrence of 'missing' elements 43 and 61 in nature (Chapter 3), the Oklo Phenomenon (Chapter 4), the occurrence of ^{244}Pu in the early solar system (Chapter 6), and isotopic anomalies in the early solar system (Chapter 7). In the present monograph, an attempt has been made to review the results from these and related investigations concerning the origin of the elements.

2.
Abundance of the Elements

"If, despite Mendeléeff's recent demurrer we assume that the elements have been evolved from one primordial form of matter, their relative abundance becomes suggestive."

F.W. Clarke, Bull. Phil. Soc. Washington 11, 131 (1889)

The idea that the elements are made of the primary matter or *prote hyle* dates back to the Greek philosopher Aristotle. Robert Boyle, who lived in 17th century England, was of the same opinion. He said: "It seems not absurd to conceive that at the first production of mixed bodies, the *universal matter* whereof among other parts of the universe consisted, was actually divided into little particles of several sizes and shapes variously moved" (Robert Boyle, "The Sceptical Chymist," 1661). Aristotle's concept of *prote hyle* was proven to be correct more than two thousand years later by Francis W. Aston's enunciation of the whole-number rule: the atomic weights of the elements are the multiples of the weight of a *proton*. Man's curiosity over the origin and nature of the elements gave rise to two major branches of 20th century science: Geochemistry and Nuclear Chemistry.

2.1. Mendeléeff and the Periodic Law

Dmitri Mendeléeff, who in 1869 promulgated the periodic law, is said to have stated [2.1]: "The periodic law, based as it is on the solid and wholesome ground of experimental research, has been evolved independently of any conception as to the nature of the elements. It does not in the least originate in the idea of an unique matter, and it has no historical connection with the relic of the torments of classical thought; and therefore it affords no more indication of the unity of matter or of the compound nature of the elements than do the laws of Avogadro and Gerhardt, or the law of specific heats, or even the conclu-

---

2. 1   Sir Edward Thorpe, History of Chemistry, Watts & Co., London, 1914, Volume II, p. 85

sions of spectrum analysis. None of the advocates of an unique matter has ever tried to explain the law from the standpoint of ideas taken from a remote antiquity, when it was found convenient to admit the existence of many gods — and of a unique matter."

A young American chemist named F.W. Clarke [2.2], however, stated in 1873: "It is probable that the chemical elements were originally developed by a process of evolution from much simpler forms of matter, as is indicated by the progressive chemical complexity observed in passing from the nebulae through the hot stars to the cold planets." Sixteen years later, in 1889, Clarke [2.3] attempted to prove his hypothesis by representing the relative abundance of the elements by a curve, taking their atomic weights for one set of ordinates. He had hoped that some sort of periodicity might be evident, but no such regularity appeared and no definite connection with the periodic law seemed to be traceable. Yet, certain other regularities were worth noticing: all of the abundant elements were low in the scale of atomic weight, reaching a maximum at 56 in iron. Above 56 the elements were comparatively rare, and only two of them, barium and strontium, appeared in Clarke's estimates.

Clarke drew the following conclusion from these observations: "If, despite Mende-léeff's recent demurrer we assume that the elements have been evolved from one primordial form of matter, their relative abundance becomes suggestive. Starting from the original *protyle*, as Crookes has called it, the process of evolution seems to have gone on slowly until oxygen was reached. At that point the process exhibited its maximum energy, and beyond it the elements forming stable oxides were the most rapidly developed, and in the largest amounts. On this supposition the scarcity of the elements above iron becomes somewhat intelligible, but the theory does not account for everything and is to be regarded as merely tentative."

2.2. The Ideas of Crookes

In his opening presidential address to the Chemical Section of the British Association at Birmingham, England, in 1886, Sir William Crookes [2.4] chose the subject of the nature and the probable, or at least possible, origin of the "so-called" elements. In this lecture, Crookes made the following remarkable statements: "I conceive, therefore, that when we say the atomic weight of, for instance, calcium is 40, we really express the fact that, while the majority of calcium atoms have an actual atomic weight of 40, there are not a few which are represented by 39 or 41, a less number by 38 and 42, and so on. We are here reminded of Newton's 'old-worn particles'. Is it not possible, or even feasible, that these heavier and lighter atoms may have been in some cases subsequently sorted out by a process resembling chemical fractionation? This sorting out may have taken place in part while atomic matter was condensing from the primal state of intense ignition, but also it may have been partly effected in geological ages by successive solutions and precipitations of the various earths. This may seem an audacious speculation, but I do not think it is beyond the power of chemistry to test its feasibility."

2. 2   F.W. Clarke, Popular Science Monthly, January 1873 issue
2. 3   F.W. Clarke, Bull. Phil. Soc. Washington 11:131 (1889)
2. 4   William Crookes, Nature 34, September 2:423 (1886)

In regard to the question concerning the origin of the elements, Crookes expressed the following opinion: "We have noticed the revolt of many leading physicists and chemists against the ordinary acceptation of the term element. We have weighed the improbability of their eternal self-existence, or their origination by chance. As a remaining alternative we have suggested their origin by a process of evolution like that of the heavenly bodies according to Laplace, and the plants and animals of our globe according to Lamarck, Darwin, and Wallace. In the general array of the elements, as known to us, we have seen a striking approximation to that of the organic world. Summing up all the above considerations we cannot, indeed, venture to assert positively that our so-called elements have been evolved from one primordial matter; but we may contend that the balance of evidence, I think, fairly weighs in favor of this speculation. This, then, is the intricate question which I have striven to unfold before you, a question that I especially commend to the young generation of chemists, not only as the most interesting, but most profoundly important, in the entire compass of our science."

2.3. Richards and Atomic Weights

Theodore W. Richards [2.5], the first American chemist to receive the Nobel Prize in chemistry, stated in 1919: "If our inconceivably ancient Universe even had any beginning, the conditions determining that beginning must even now be engraved in the atomic weights. They are the hieroglyphics which tell in a language of their own the story of the birth or evolution of all matter, and the Periodic Table containing the classification of the elements is the Rosetta Stone, which may enable us to interpret them. Until, however, these hieroglyphics are clearly visible in their true form, we can not hope for an interpretation. The first task of the investigator is to define sharply the outlines of these graven characters, in order that their true form may be manifest. Then perhaps there is hope of deciphering their meaning."

According to Richards, the subject of atomic weights in 1919 was far from being a completed and closed chapter of Science. He felt that the future opened up a prospect of almost endless further investigations and the work completed by him was thus only a beginning. He thus noted: "Each generation builds upon the results of its predecessors. Stas improved the admirable work of the master Berzelius, using many of his methods, with improved appliances and wider chemical knowledge of the later date. As you have just seen, the more recent researches have improved upon those of Stas. In years to come, let us hope that yet finer means of research and yet deeper chemical knowledge may make possible further improvements, yielding for mankind a more profound and far-reaching knowledge of the secrets of the wonderful Universe in which our lot is cast."

There are three places in the Periodic Table, where the order of the atomic number and the atomic weight is reversed. The anomalies occur in the sequences: Ar-K-Ca, Fe-Co-Ni and Te-I-Xe. These anomalies could not be explained in the days of Richards, but in 1919 Francis W. Aston constructed the first mass spectrograph, which enabled one to determine the relative abundances of the stable isotopes of the elements. It thus became possible to measure quite accurately not only the "average" weights of the elements, but also the exact isotopic compositions. Modern versions of Aston's mass spectrograph are

---

2. 5   Theodore W. Richards, Atomic Weights, Nobel Lecture, December 6, 1919

now being used by many researchers in their exploration into the origin of the elements in the Universe.

Francis W. Aston, born on September 1, 1877, at Harborne, Birmingham, England, and Frederick Soddy, born on September 2, 1877, at Eastbourne, Sussex, England, proved that substances could exist with identical, or practically identical, chemical and spectroscopic properties but different atomic weights.

The need for specific names for such substances soon became imperative, and Soddy suggested the word *Isotopes* ($\iota\sigma o\varsigma$, equal, $\tau o\pi o\varsigma$, place) because they occupied the same place in the periodic table of the elements. Soddy received the Nobel Prize in chemistry in 1921 for his contributions to our knowledge of the chemistry of radioactive substances, and his investigations into the origin and nature of isotopes. Aston received the Nobel Prize in chemistry in 1922 for his discovery, by means of his mass spectrograph, of isotopes in a large number of non-radioactive elements, and for his enunciation of the whole-number rule.

2.4. Clarke's Numbers

In his attempt to establish the relative abundance of the elements, Clarke [2.6] noted that we must bear in mind the limitations of our experience. The knowledge of terrestrial matter extended but a short distance below the surface of the earth, and beyond that we can only indulge in speculation. The atmosphere, the ocean and a thin shell of solids were all that could be examined in the days of Clarke. For the first two layers the information was reasonably good and their masses were approximately known, but some arbitrary limit had to be assumed for the last one. Clarke noted that it seemed probable to a depth of 10 miles below sea level the rocky material could not vary greatly from the volcanic outflows which we recognize at the surface, and this thickness gave us a quantitative basis for study.

Clarke published a book entitled "The Data of Geochemistry" in 1908 as Bulletin 330 of the U.S. Geological Survey. The fifth and enlarged edition of this book was published as Bulletin 770 in 1924 [2.6]. The average composition of known terrestrial matter calculated by Clarke is given in this book. The so-called Clarke's numbers were obtained primarily on the basis of a compilation of rock analyses by H.S. Washington, in which 8,602 analyses of igneous rocks were brought together and divided into groups according to their quality. Of these only those reported as "superior", 5,159 in number, have been used in the computations.

2.5. The Rule of Harkins

In 1917, W.D. Harkins [2.7] published a paper entitled "The Evolution of the Elements and the Stability of Complex Atoms. I. A New Periodic System of the Elements which shows a Relation Between the Abundance of the Elements and the Structure of the Nuclei of Atoms". He stated: "In studying the relative abundance of the elements the ideal

2. 6   F.W. Clarke, The Data of Geochemistry, Washington, Government Printing Office, 1924
2. 7   W.D. Harkins, J. Am. Chem. Soc. 39:856 (1917)

method would be to sample one or more solar systems at the desired stage of evolution, and to make a quantitative analysis for all the 92 elements of the ordinary (periodic) system. Since this is impossible, even in the case of the earth, it might be considered that sufficiently good data could be obtained from the earth's crust, or the lithosphere. However, there are several important factors which cause our knowledge of the quantitative composition of the earth's crust to be of much less value for the solution of our problem than it might seem to possess on first thought."

After pointing out the difficulties involved in attempting to study the surface of the sun by the spectroscopic method, Harkins went on to make the following important statement: "There is, however, material available of which accurate quantitative analyses can be made, and which falls upon the earth's surface from space. The bodies which fall are called *meteorites*, and no matter what theory of their origin is accepted, it is evident that this material comes from much more varied sources than the rocks on the surface of the earth. In any event, it seems probable that the meteorites represent more accurately the average composition of material at the stage of evolution corresponding to the earth than does the very limited part of the earth's material to which we have access."

Based on the analytical data available then, the average values of 318 iron and 125 stone meteorites, 443 in all, Harkins showed that the first seven elements in order of abundance were O, Fe, Ni, Si, Mg, S, and Ca: all of them having *even* atomic numbers and in addition they made up 98.6 percent of the material of the meteorites. When the abundances of the elements in stony meteorites were plotted against the atomic numbers, the following pattern emerged at once: every even-numbered element is more abundant than the two adjacent odd-numbered elements. This important relationship is called the rule of Harkins, or Oddo-Harkins, since in 1914, Oddo [2.8] reported that elements whose atomic weights are divisible by 4 made up for the bulk (86.5 percent) of the mass of the upper lithosphere.

Since the surface of the earth has been subjected to far-reaching differentiative processes, Harkins felt that it would be valuable if evidence could be obtained in regard to elements which have not been thus affected. Any group of elements, which are very much alike both chemically and in physical properties, would be affected more nearly to the same extent than those which differ widely. The elements most nearly alike are the rare earths. Harkins therefore made an estimate of the relative abundance of the members of the rare earth group, and then obtained estimates from two celebrated workers in this special field (Harkins did not mention their names in his 1917 paper). The three independent estimates agreed for all of the rare earths included in Table 2.1., where the letter c indicates common and r indicates rare; cc represents very common and so on. The results from the analysis by the X-ray method carried out by E. Minami [2.9] in 1935 in V.M. Goldschmidt's laboratories in Göttingen are also shown in Table 2.1 for comparison.

The rule of Harkins generally holds not only for the elements, but also for the isotopes. It is interesting to note, however, that there is a notable exception. The isotopic composition of xenon was first measured by Aston [2.10] in 1930, as shown in Table 2.2.

2. 8 G. Oddo, Z. Anorg. Chem. 87:253 (1914)
2. 9 E. Minami, Nachr. Ges. Wiss. Göttingen IV, N. F. 1, No. 14:155 (1935)
2.10 F.W. Aston, Proc. Roy. Soc. A. 126:511 (1930); Mass-Spectra and Isotopes, London, 1933, p. 107.

Table 2.1. The rule of Harkins and the abundance of the rare earth elements

Z	Element	Abundance estimated by Harkins (1917)	Average abundance in shales from Europe and Japan (E. Minami, 1935)
57	La	*c*	21.6
58	Ce	*cc*	56.6
59	Pr	*r*	6.57
60	Nd	*c*	27.8
61	–	*rrr*	–
62	Sm	*c*	7.50
63	Eu	*rr*	1.23
64	Gd	*r*	7.33
65	Tb	*rrr*	1.05
66	Dy	–	5.13
67	Ho	–	1.39
68	Er	–	2.83
69	Tm	–	0.23
70	Yb	–	3.03
71	Lu	–	0.75

Table 2.2. Isotopic composition of xenon measured by Aston in 1930

Mass number	124	126	128	129	130	131	132	134	136
Abundance	0.08	0.08	2.30	27.13	4.18	20.67	26.45	10.31	8.79

According to the early measurements by Aston, the most abundant isotope of xenon is ^{129}Xe with an odd mass number and its abundance is far greater than those of the neighboring even mass number isotopes ^{128}Xe and ^{130}Xe. A more detailed discussion on these isotope anomalies in xenon will be presented in Chapters 6 and 7.

2.6. The 1930 Estimates by Noddack

Ida and Walter Noddack [2.11] began their search for two "missing" elements 43 and 75 in 1922. The concentrations of these elements in minerals were extremely low and it was necessary to enrich them by physical or chemical means. During the process of enrichment, the existence of a series of other rare elements, which were also present in various minerals at extremely low concentrations, also became apparent. Thus the Noddacks extended their investigation into the study of the abundance of a large number of rare elements. They used the X-ray spectroscopic method as the major analytical tool, as well as the ordinary spectroscopic method combined with the chemical enrichment procedure. Some 1,600 terrestrial and extraterrestrial samples were analyzed, as well as mixtures of many representative samples.

2.11   I. and W. Noddack, Naturwissenschaften 18:757 (1930).

Out of 42 stone meteorites, which showed the least signs of weathering, all the molten crusts were removed. The samples were crushed and the metals were removed by a magnet. The non-magnetic silicates were also studied separately. For the analysis of iron phase, 16 iron meteorites were used. All the visible silicates and troilites (FeS) were removed from the sample. For the analyses of sulfide phase, troilites were separated from 5 iron meteorites. The analyses were carried out for all the elements known at that time with the exception of hydrogen, rare gases and the short-lived radioactive elements.

In calculating the average relative abundance of the elements (H) in the meteorites, a serious difficulty is encountered; namely, the ratio of stone : iron : troilite can not be determined easily. Assuming the average troilite content of the stone meteorite to be 5.5 percent and comparing the densities of the 3 different kinds of meteorites with the average density (= 5.1) of the four inner planets and the Moon, the Noddacks accepted the ratio of

stone : iron : troilite = 1 : 0.68 : 0.098.

Fig. 2.1 shows the cosmic abundance of the elements obtained by the Noddacks in 1930. It should be noted that the abundance distribution shown here represents only the inner part of the solar system, excluding the Sun. Little is known of the chemical nature of the outer planets. Their low densities (for example, Jupiter = 1.3) suggested that at the formation of the solar system a fractionation of the chemical elements must have taken place, so that the outer region contains less iron, but more rare gases.

The curve beautifully demonstrated the validity of the rule of Harkins, but there seemed to be an unexpected new periodicity represented by the maxima for Si, Sn, and Pb, and the minima for Sc, Ga, In, Tl, Cl, Br, I, element 43 (which the Noddacks called "Masurium", Ma) and Re. They stated that the periodicity of the abundance distribution supported the view that the nucleus has a structure, which is similar to the fillings of the orbital electrons.

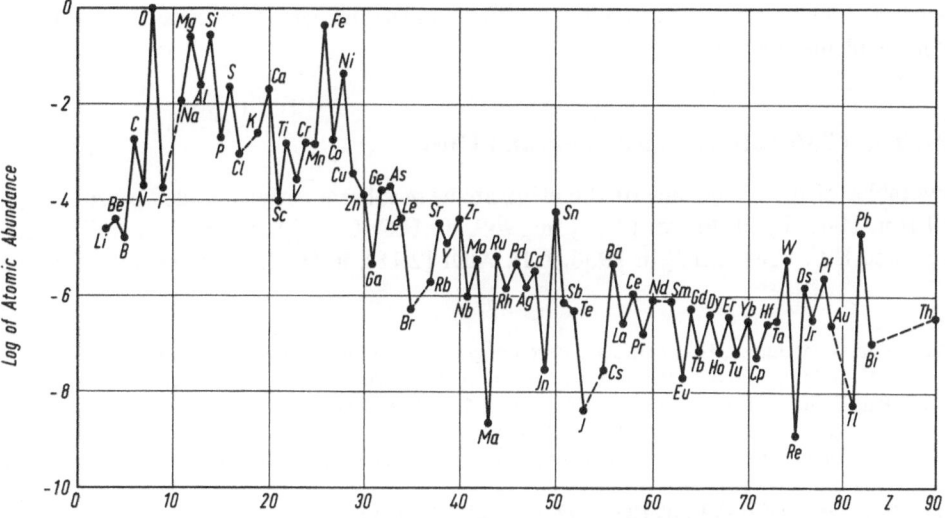

Fig. 2.1. Cosmic abundance of the elements according to Ida and Walter Noddack (Naturwissenschaften 18:757 (1930)). The abundance of oxygen is taken to be equal to unit

2.7. The 1938 Estimates by Goldschmidt

In 1938, V.M. Goldschmidt [2.12] compiled a new table of the abundance of the elements. The revised version of this table was given in Goldschmidt's book "Geochemistry" [2.13] edited by Alex Muir, which appeared in 1954. Goldschmidt's abundance table was extensively studied by subsequent investigators. Appendix I shows the abundance data given by Goldschmidt, as they appeared in the 1954 edition of "Geochemistry".

2.8. Geochemical Classification of the Elements

Goldschmidt [2.12, 2.13] stated that, based on the existing data on the affinity of various elements for oxygen and sulfur, we can obtain a geochemical classification of the elements. From the free energy of formation of oxides, combined with the free energy of formation of various elements with iron, he drew up a list of elements which are concentrated in the iron phase of meteorites, and probably also in the supposed iron core of the earth. He called these elements *siderophil* elements, or elements tending to concentrate in metallic iron. Typical examples are nickel, cobalt, and the metals of the palladium and platinum groups.

A second group is formed by those elements which have a greater free energy of oxidation, per gram atom of oxygen, than iron. To these elements, Goldschmidt gave the name *lithophil* elements, or elements tending to concentrate in stony matter. They concentrate in the stony matter of the earth, as well as in the stony matter of meteorites, as oxides or silicates.

A third group is formed by elements concentrated in the sulfide phases of meteorites, such as troilite (FeS) and Goldschmidt called them *chalcophil* elements. Typical examples are: S, Se, Te, As, Sb, Bi, Ga, Pb, Zn, Cd, Hg, Cu, Ag, etc. Elements which occur either in the uncombined state, such as oxygen, nitrogen and rare gases, etc., or as volatile compounds, e.g., carbon as CO and CO_2, or boron, probably as B_2O_3 or BCl_3, will concentrate in the gaseous primordial atmosphere. According to Goldschmidt, they are *atmophil* elements. Finally, we can distinguish the *biophil* elements which are concentrated in and by living plants and animals.

2.9. The 1956 Estimates by Suess and Urey

New tables of the abundance of the elements were subsequently compiled by H.E. Suess [2.14] in 1947, by H. Brown [2.15] in 1949, by H.C. Urey [2.16] in 1952 and by H.E. Suess and H.C. Urey [2.17] in 1956. L.H. Aller [2.18], in 1961, compiled a table of ato-

2.12   V.M. Goldschmidt, Geochemische Verteilungsgesetze IX, Videnskakademien, Oslo, 1938
2.13   V.M. Goldschmidt, Geochemistry, edited by A. Muir, Clarendon Press, Oxford, 1954
2.14   Hans E. Suess, Z. Naturforsch. 2a:311−321, 604−608 (1947)
2.15   H. Brown, Rev. Mod. Phys. 21:625 (1949)
2.16   H.C. Urey, The Planets: Their Origin and Development, Yale University Press, New Haven, 1952
2.17   H.E. Suess and H.C. Urey, Rev. Mod. Phys. 28:53 (1956)
2.18   L.H. Aller, The Abundance of the Elements, Interscience Monographs and Texts in Physics and Astronomy, Vol. VII, Interscience Publishers, New York, 1961

mic abundances for the primordial solar system based primarily on astrophysical data. A.G.W. Cameron [2.19], in 1967, prepared a new abundance table based as much as possible on measurements in Type I carbonaceous chondrites. Appendix II shows the 1956 Suess-Urey cosmic abundance values for the individual nuclei. Note that the abundance values are given here relative to Si = 10^6. Plots of logarithms of the abundances H against mass numbers A are shown in Fig. 2.2.

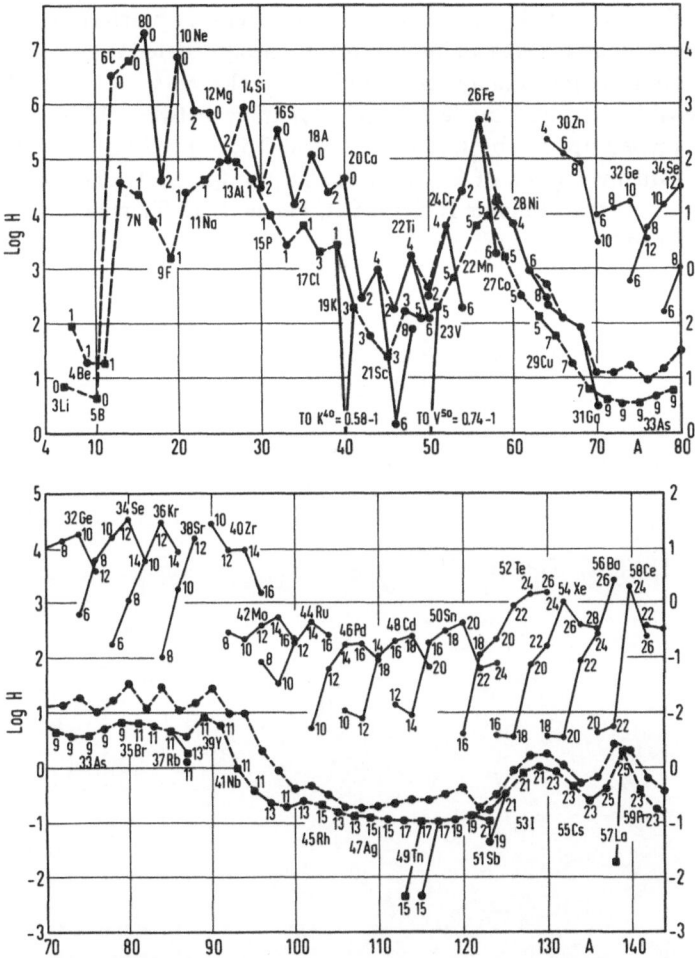

Fig. 2.2. Logarithm of abundance (silicon = 10^6) plotted against mass number (A) according to Suess and Urey (1956).

The even and odd mass numbers are on separate curves. The neutron excess numbers (I) are shown at each point. The curve without I indicated, shows the sum of the isobaric abundances for the even A series. Note that the right-hand scale is for the curve representing the even A series (light lines) beginning with A = 64 (Zn)

2.19   A.G.W. Cameron, Origin and Distribution of the Elements, edited by L.H. Ahrens, Pergamon Press, 1968, p. 125

The 1956 tabulation of the cosmic abundance of the elements by Suess and Urey laid the foundation for the B_2FH theory of the element synthesis in stars published in 1957 by Burbidge, Burbidge, Fowler and Hoyle [2.20]. The main features of the Suess-Urey abundance table will be discussed in *Chapter 5. Synthesis of the Elements in Stars.*

In 1975, Trimble [2.21] published a review article entitled "The Origin and Abundances of the Chemical Elements" and compiled a table of relative abundances by number of the chemical elements in the meteorites, solar photosphere, solar corona, and galactic cosmic ray sources (see Appendix III). She noted that the current standard abundance distributions to which theoretical predictions are usually compared are those by Cameron [2.19, 2.22] and the latter of these, which does not differ very much from the former, is not regarded by everyone as being an improvement. She also stated that in fact the standard abundances have been remarkably stable and only 8 elements out of the 81 tabulated had abundances which changed by factors of 3 or more between 1956 and 1968: C, Ga, and Pb were revised upward by factors 3, 4, and 8.5, respectively, and Be, La, Sm, Ta, and W downward by factors of 25, 4, 3, 3, and 3, respectively. The most recent table of the abundance of the elements is that of H. Palme, H.E. Suess and H.D. Zeh [2.23], which was published in 1981.

2.20   E.M. Burbidge, G.R. Burbidge, W.A. Fowler and F. Hoyle, Rev. Mod. Phys. 29:547 (1957)
2.21   V. Trimble, Rev. Mod. Phys. 47:877 (1975)
2.22   A.G.W. Cameron, Space Sci. Rev. 15:121 (1973)
2.23   H. Palme, H.E. Suess and H.D. Zeh, Astronomy and Astrophysics, Table 4. In: Landolt-Börnstein Numerical Data and Functional Relationships in Science and Technology, Vol. 2, Berlin, Heidelberg, New York, 1981, p. 267

3.
Elements 43 and 61 in Nature

"Alle chemischen Elemente kommen in allen Mineralien vor."

Ida Noddack, Angewandte Chemie 47, 835 (1936)

Element 43, technetium, was discovered in certain stars and a renewed search for this element in rocks and minerals began in the early 1950's. The results obtained during the 1960's indicated, however, that the terrestrial abundances of the man-made elements such as technetium and promethium are essentially controlled by their production through spontaneous fission of ^{238}U. The elements 43 and 61 occur in the earth's crust in the same sense as, for example, radium and polonium occur in the lithosphere in radioactive equilibria with their long-lived precursor ^{238}U.

3.1. The All-Present Theory of Noddack

In 1936, Ida Noddack [3.1] published a paper entitled "Über die Allgegenwart der chemischen Elemente", in which she reported that detailed analysis of a sample of zinkblende revealed the presence of all except 16 elements out of 89 known elements. These 16 elements were rare gases, rare alkali elements, and Be, B, N, F, I, Ma (element 43), Nb and Ta. She felt confident that these remaining 16 elements were present in minerals and their presence could be demonstrated, if a suitable enrichment procedure was used for each element.

As a second example, she reported the result of a chemical analysis of a sample of copper shale (Kupferschiefer) from Mansfeld (see Table 3.1). Here one also finds the same picture: most of the known elements are detected in this mineral, and the remaining ele-

---

3. 1   Ida Noddack, Angewandte Chemie 47:835 (1936)

Table 3.1. Chemical Composition of Copper Shale from Mansfeld, according to Ida Noddack (1936)

I	O = 42% Si = 26%	Al = 9% Ca = 6% Mg = 4%	Fe = 4% Cu = 3% S = 6%		
II	Na = 1.8% K = 0.4% C = 0.9%	Sr = 0.2% Cl = 0.6%	Ni = 0.3% Zn = 0.5% Mn = 0.2%	Ti = 0.3% N = 0.2% V = 0.1%	
III	Mo = 0.08% Cr = 0.04% F = 0.02%	Sr = 0.04% Ba = 0.02% Zr = 0.02%	Co = 0.05% As = 0.09% Pb = 0.04%	Se = 0.03% Sn = 0.02% Hf = 0.002%	Br = 0.004% Li = 0.002% B = 0.003%
IV	$Li = 8 \cdot 10^{-6}$ $W = 2 \cdot 10^{-6}$ $Ag = 3 \cdot 10^{-6}$	$Ga = 9 \cdot 10^{-6}$ $Cd = 8 \cdot 10^{-6}$ $Th = 8 \cdot 10^{-6}$	$Ge = 8 \cdot 10^{-6}$ $Bi = 5 \cdot 10^{-6}$ $Pd = 2 \cdot 10^{-6}$	$Te = 2 \cdot 10^{-6}$ $Sb = 5 \cdot 10^{-6}$ $Pt = 5 \cdot 10^{-6}$	$Y = 3 \cdot 10^{-6}$ $La = 2 \cdot 10^{-6}$ $Sc = 4 \cdot 10^{-6}$
V	$Ru = 3 \cdot 10^{-7}$ $Au = 3 \cdot 10^{-7}$ $U = 2 \cdot 10^{-7}$ $Be = 5 \cdot 10^{-7}$	$In = 3 \cdot 10^{-7}$ $Tl = 2 \cdot 10^{-7}$ $Nb = 8 \cdot 10^{-7}$ $Ta = 6 \cdot 10^{-7}$	$Re = 4 \cdot 10^{-7}$ $Ir = 4 \cdot 10^{-8}$ $J = 4 \cdot 10^{-8}$	$Os = 3 \cdot 10^{-8}$ $Rh = 2 \cdot 10^{-8}$ $Hg = 2 \cdot 10^{-8}$	

ments were not sought after hard enough. The general impression was that through sufficient labor and time, one should be able to find all the remaining elements in this mineral.

The analyses of not only the natural minerals, but also organic substances and artificial products reveal the same trend: the greater the detection sensitivity, the greater the number of chemical elements detected in the sample. Thus, one obtains an impression that if the analytical sensitivity is sufficiently increased, *all the elements may be detected in all the substances.*

Noddack was obviously searching for the "missing" elements 43 (masurium) and 61 (illinium) [3.2]. These two elements were never found in terrestrial minerals and all the efforts to enrich these elements ended in failure. The reason why these elements were absent in nature was that they were radioactive and the half-lives of the isotopes of these elements were much shorter than the age of the earth. All the isotopes of these elements originally synthesized in stars had long since decayed away. It does not necessarily follow, however, that Noddack made an error in postulating that all the elements may be detected in all the substances. The copper shale from Mansfeld contains a small quantity of uranium as shown in Table 3.1. As we shall see later, elements 43 and 61 do occur in minerals or ores containing uranium, in the same sense as, for example, radium and polonium occur in rocks and minerals in radioactive equilibria with their long-lived precursor ^{238}U.

3.2. Discoveries of Elements 43 and 61 by Artificial Means

In 1937, Perrier and Segrè [3.3] found that molybdenum, when irradiated with deuterons in the Berkeley cyclotron, exhibited a strong unknown radioactivity. A careful study of

3. 2   J.A. Harris and B.C. Hopkins, J. Am. Chem. Soc. 48:1585 (1926)
3. 3   C. Perrier and E. Segrè, J. Chem. Phys. 5:712 (1937); 7:155 (1939); Nature 140:193 (1937)

this new radioactivity led to the conclusion that element 43 is formed by the nuclear reactions:

$$^{94}Mo\ (d,n)\ ^{95}43\ (60\text{-day half-life})$$

and

$$^{96}Mo\ (d,\ n)\ ^{97}43\ (90\text{-day half-life}).$$

With unweighable amounts, observations were made on the chemical behavior of element 43. Perrier and Segrè, in 1947, proposed the name technetium, with the symbol Tc. The name was derived from a Greek word meaning "artificial", which is appropriate since this element was the first new element to be discovered by means of artificial synthesis.

In 1947, Marinsky, Glendenin and Coryell [3.4] reported the first positive chemical identification of isotopes of element 61. They studied elution characteristics of rare earth ions from Amberlite IR-1 absorption columns using buffered citrate solutions and found that the order of elution was the reverse of atomic number with yttrium falling near gadolinium. Column separations on fission product mixtures containing activities of praseodymium, neodymium, and element 61 led to the first chemical identification of the new element. The two new isotopes which they discovered were 2.6-year ^{147}Pm and 53-hour ^{149}Pm.

3.3. Magic Numbers

Elsasser [3.5], in 1933 and 1934, made the first suggestion that certain numbers of neutrons or protons in the nucleus form a particularly stable configuration. The complete evidence for this was not summarized until 1948, however, when Maria G. Mayer [3.6] reported that nuclei with 20, 50, 82 or 126 neutrons or protons are particularly stable. In 1949 and 1950, the nuclear shell model [3.7, 3.8, 3.9] was proposed to explain the fact that at these "magic numbers" there occurs something analogous to the closure of shells in the electronic structure of the noble gas atoms. The binding energy of nuclei can be quite well represented by a continuous formula, but some irregularities are found, which are associated with neutron numbers $N = 2, 8, 20, 28, 50, 82$, and 126, and with the same proton numbers. A few indications of the significance of the magic numbers will be presented here.

In the case of Sn ($Z = 50$), this element has the greatest number of isotopes of any element, namely, 10. Its heaviest and lightest nuclei differ by 12 neutrons:

$$^{112}Sn,\ ^{114}Sn,\ ^{115}Sn,\ ^{116}Sn,\ ^{117}Sn,\ ^{118}Sn,\ ^{119}Sn,\ ^{120}Sn,\ ^{122}Sn,\ \text{and}\ ^{124}Sn.$$

3. 4   J.A. Marinsky, L.E. Glendenin and C.D. Coryell, J. Am. Chem. Soc. 69:2781 (1947)
3. 5   W. Elsasser, J. de phys. et rad. 4:549 (1933); 5:625 (1934)
3. 6   M.G. Mayer, Phys. Rev. 74:235 (1948)
3. 7   M.G. Mayer and J.H.D. Jensen, Elementary Theory of Nuclear Shell Structure, John Wiley & Sons, New York, 1955
3. 8   M.G. Mayer, Phys. Rev. 75:1969 (1949); 78:16, 22 (1950)
3. 9   O. Haxel, J.H.D. Jensen and H.E. Suess, Phys. Rev. 75:1766 (1949); Z. Physik 128:295 (1950)

A spread of 12 mass units occurs in only one other case, namely Xe, which has 9 stable isotopes:

$$^{124}Xe, \ ^{126}Xe, \ ^{128}Xe, \ ^{129}Xe, \ ^{130}Xe, \ ^{131}Xe, \ ^{132}Xe, \ ^{134}Xe, \text{ and } ^{136}Xe.$$

According to Mayer, this may be attributed to the extra-stability of ^{136}Xe with 82 neutrons. The next largest difference (10 mass unit) occurs once only at Sm and this is again due to the extra-stability of ^{144}Sm with 82 neutrons.

In the case of Ca (Z = 20), the difference in mass number between the lightest and heaviest isotope is eight:

$$^{40}Ca, \ ^{42}Ca, \ ^{43}Ca, \ ^{44}Ca, \ ^{46}Ca, \text{ and } ^{48}Ca.$$

Note that the isotopes ^{40}Ca and ^{48}Ca are "doubly-magic": the former contains 20 protons and 20 neutrons and the latter 20 protons and 28 neutrons.

The semi-empirical formula for the mass of an atom with mass number A and charge Z is [3.10, 3.11]:

$$M_{A,Z} = A - 0.00081Z - 0.00611A + 0.014 \, A^{2/3} + 0.083 \, (A/2 - Z)^2 \, A^{-1}$$
$$- 0.000627 \, Z^2 \, A^{-1/3} + \delta \tag{3.1}$$

where,

$$
\begin{aligned}
\delta \ &= 0 &&\text{for odd A,} \\
&= -\ 0.036 \, A^{-3/2} &&\text{for A even Z even} \\
&= +\ 0.036 \, A^{-3/2} &&\text{for A odd Z odd}
\end{aligned}
\tag{3.2}
$$

According to equation (3.1), the isotope

$$^{136}_{54}Xe_{82}$$

should be unstable against β decay by about 2 MeV, but actually it is stable. ^{140}Ba should be stable and ^{144}Sm should be unstable against K-capture by 0.6 MeV, also according to equation (3.1), but actually

$$^{144}_{62}Sm_{82}$$

is stable.

It is interesting to note here that it was quite fortunate for geochemists and cosmo-chemists that ^{136}Xe turned out to be a stable isotope, since otherwise it would have made it quite difficult to study the isotopic compositions of the xenon produced by spontanous fission of ^{238}U and ^{244}Pu in terrestrial rocks and meteorites. A more detailed discussion on the subject of xenon isotope anomalies will be presented in Chapters 6 and 7.

3.10   N. Bohr and J.A. Wheeler, Phys. Rev. 56:426 (1939)

3.11   G.B. von Albada, Astrophys. J. 105:393 (1947)

In 1951, Kowarski [3.12] reported that the abnormal instability of *all* isotopes of elements 43 (Tc) and 61 (Pm) may have something to do with the occurrence, close by, of magic numbers 50 and 82. Suess [3.13], also in 1951, reported that in order to find an explanation for the absence of beta-stable isotopes of the elements Tc and Pm, a more general abnormality in the region following the closing of the 50 and 82 neutron shells should be taken into consideration.

3.4. Technetium in Stars

The important discovery of technetium in certain stars was announced by P.W. Merrill [3.14] in 1952. The spectrum of the *artificial* element technetium was thoroughly investigated by Meggers and Scribner in 1950 at the National Bureau of Standards, and this work made astronomical investigations possible. Spectrograms, dispersion 10A/mm, taken by Merrill and others taken by I.S. Bowen with the 200-inch telescope, showed several lines of neutral technetium in the spectra of S-type stars, especially of long-lived variables. The strongest of these lines were analogous to the well-known triplet at λ 4030 in the spectrum of manganese. Stellar spectra of type S were also characterized by bands of zirconium oxide and by relatively strong lines of heavy metals such as Zr and Ba.

Merrill stated that it was surprising to find an unstable element like Tc in the stars. This must be interpreted as either (a) a stable unknown isotope of technetium actually exists although not yet found on earth; or (b) S-type stars somehow produce technetium as they go along; or (c) S-type stars represent a comparatively transient phase of stellar existence.

3.5. Long-Lived Isotopes of Technetium

In 1955, Boyd and his co-workers [3.15] at Oak Ridge National Laboratory reported on the production and identification of long-lived technetium isotopes at masses 97, 98 and 99. At that time, it appeared that the technetium isotopes at masses 97 and 99 were unstable and decayed with half-lives of less than 10^6 years and that ^{98}Tc was a "missing" isotope.

A large piece of molybdenum metal was bombarded with 22-MeV protons from the ORNL 86-inch cyclotron for a time sufficient to produce about 10 micrograms of ^{98}Tc. The technetium fraction was separated and purified by anion exchange chromatography and approximately one microgram of technetium was placed on an iridium strip employed as a filament source in a $60°$ mass spectrometer of eight inch radius. Prominent peaks at masses 97, 98 and 99 were observed. Boyd et al. [3.15] noted that the results were of interest for a number of reasons: (a) weighable quantities of three long-lived technetium isotopes were seen on a mass spectrometer for the first time; (b) the "missing" isotope,

3.12   L. Kowarski, Phys. Rev. 78:477 (1951)
3.13   H.E. Suess, Phys. Rev 81:1071 (1951)
3.14   P.W. Merrill, Science 115:484 (1952)
3.15   G.E. Boyd, J.R. Sites, Q.V. Larson and C.R. Baldock, Phys. Rev. 99:1030 (1955)

[98]Tc was observed and hence must be long-lived; and (c) the occurrence of a long-lived isomer of [100]Tc seemed to be excluded. Fig. 3.1 shows an X-ray spectrogram taken on a mixture of Mo, Tc and Ru.

3.6. Reported Discoveries of Technetium in Terrestrial Minerals

It appears that the Noddacks [3.16] believed in the natural occurrence of the element 43 (Ma) as late as in 1954. Together with O. Berg, they reported to have discovered in 1925 the elements 43 and 75 and proposed the names "Masurium" and "Rhenium". While within a few years rhenium became an easily available element, no report was published on masurium by the discoverers since 1929.

The Noddacks stated that the reason for this was an extremely low abundance of masurium on the one hand, and the lack of starting materials on the other. By 1944, some progress had been made in the enrichment of this element, but all the enriched samples were lost in the final days of the war and the work had to be interrupted for 5 years. In 1952, an enriched sample of masurium was again obtained.

Meanwhile, in 1954, W. Herr [3.17] at the Max-Planck-Institut für Chemie, Mainz, reported that it was quite likely that extremely small quantities of [98]Tc existed in certain molybdenum minerals. Chemical behavior of technetium is similar to Mn and Re, but the analytical chemistry of technetium is closer to that of Re. Because of the fact that technetium can exist in various valency states, it is difficult to predict its geochemical behavior, but it is likely that technetium will be enriched in Re-rich minerals.

Herr and co-workers [3.18, 3.19], in 1954, made the important discovery that Os in Re-rich minerals showed an isotope anomaly due to the natural radioactivity of [187]Re.

$$^{187}\text{Re} \xrightarrow{\beta^-} {}^{187}\text{Os}$$

In an attempt to find the naturally occurring technetium in minerals, Herr treated a 370-gram sample of molybdenite with fuming HNO_3 and also with $HClO_4$-distillation. A small quantity of artificial technetium was added to the sample as tracer. Re and Tc were then precipitated with tetraphenylarsonium chloride. Re was then removed through repeated distillations from $HClO_4$ and H_2S precipitation from 9N HCl. The technetium fraction, which was practically weightless, was finally obtained together with 1 mg of Cu added as carrier, and was irradiated with neutrons in the Harwell-reactor. If [98]Tc was present in the sample, 6-hour ^{99m}Tc should have been produced through the reaction

$$^{98}\text{Tc} (n, \gamma) \, {}^{99m}\text{Tc}.$$

However, if the sample contained molybdenum as an impurity, the reaction

$$^{98}\text{Mo} (n, \gamma) \, {}^{99}\text{Mo} \xrightarrow{\beta^-} {}^{99m}\text{Tc}$$

3.16   W. Noddack and I. Noddack, Angewandte Chemie 66:752 (1954)
3.17   W. Herr, Z. Naturforsch. 9a, No. 10:907 (1954)
3.18   W. Herr, H. Hintenberger and H. Voshage, Phys. Rev. 95:1691 (1954)
3.19   H. Hintenberger, W. Herr and H. Voshage, Phys. Rev. 95:1960 (1954)

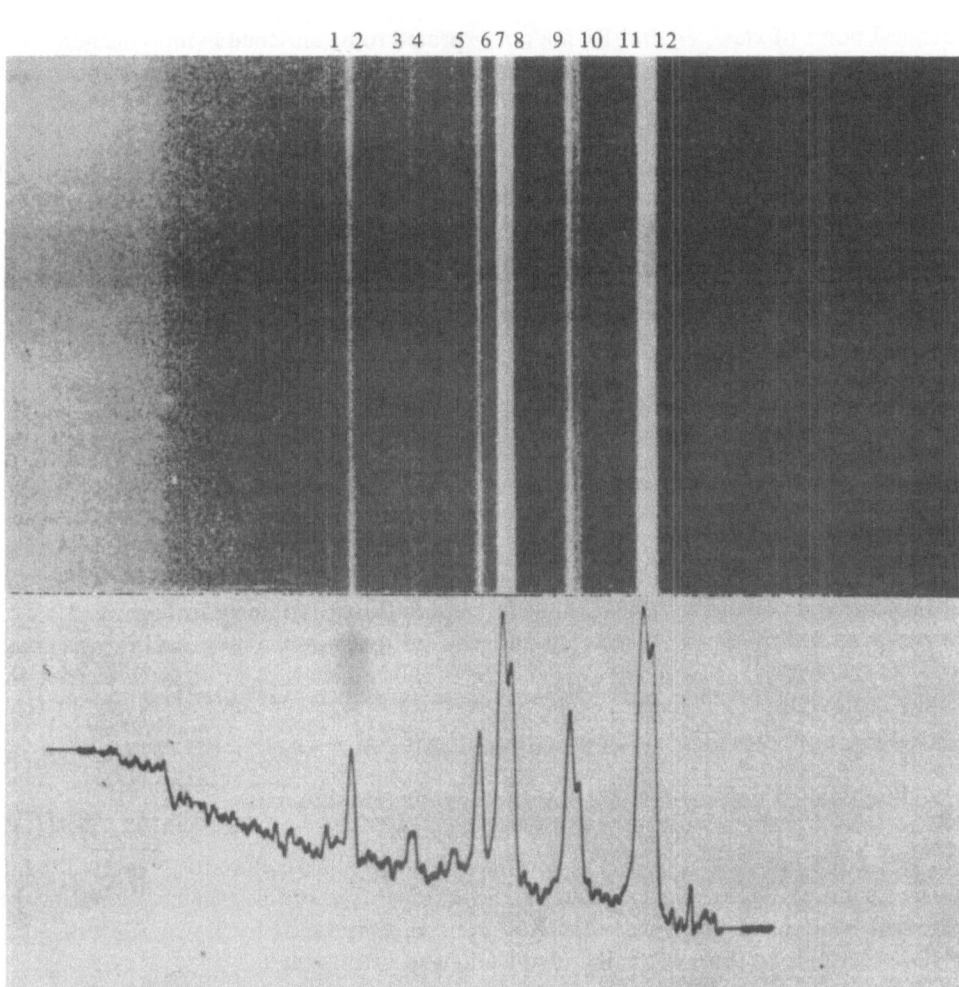

$1 - Ru\ K\beta_1$	$4 - 43\ K\beta_1$	$7 - Ru\ K\alpha_1$	$10 - 43\ K\alpha_2$
$2 - Ru\ K\beta_1$	$5 - Mo\ K\beta_2$	$8 - Ru\ K\alpha_2$	$11 - Mo\ K\alpha_1$
$3 - 43\ K\beta_2$	$6 - Mo\ K\beta_1$	$9 - 43\ K\alpha_1$	$12 - Mo\ K\alpha_2$

Fig. 3.1. X-ray spectrogram taken on a mixture of Mo, Tc, and Ru: Courtesy of Dr. G.E. Boyd of the Oak Ridge National Laboratory. This photograph is the first evidence for the isolation of weighable quantities of element 43

could also produce the 6-hour activity. Therefore, a control experiment was carried out using 1.1 mg of Re under the same experimental conditions and it was found that the 6-hour activity produced was very weak. The sample was irradiated for 9 hours in the pile and the neutron-flux was 1.2×10^{12} per cm^2 per second. A fairly strong 6-hour activity was observed, but the gross decay curve also showed the existence of 3.6-day ^{186}Re activity. Based on these experiments, Herr concluded that it was quite likely that this ore contained an extremely small quantity of ^{98}Tc.

Herr also reported that the same ore contained a relatively large quantity of Os, which turned out to be a pure ^{187}Os, as well as a very small quantity of ruthenium. From a geochemical point of view, Os and Ru are not expected to be enriched in molybdenum ores. Initial neutron-activation experiments moreover gave an indication that the isotopic composition of this ruthenium was not normal. Herr therefore suggested that ^{98}Tc might be beta-unstable and decayed to the stable ^{98}Ru.

In 1955, Alperovitch and Miller [3.20] also reported the possibility of occurrence of ^{98}Tc in various minerals. They noted that ^{99m}Tc produced by the ^{98}Tc (n, γ) reaction could be readily identified by its 6-hour half-life and 140-KeV γ-ray, provided that its production by the following three alternative processes could be ruled out:

$$^{98}\text{Mo (n, } \gamma\text{) } ^{99}\text{Mo} \longrightarrow ^{99m}\text{Tc};$$
$$^{99}\text{Tc (n, n}'\text{) } ^{99m}\text{Tc; and}$$
$$^{99}\text{Ru (n, p) } ^{99m}\text{Tc}$$

It is therefore important that the concentrates to be isolated from selected minerals be free of all traces of Mo, Ru and 2×10^5-year ^{99}Tc. A radiochemical procedure was developed by a combination of anion-exchange chromatographic and distillation techniques. Of the twelve samples examined, six gave positive results and two gave negative results; in the remaining four cases the identification was uncertain.

Anders and co-workers [3.21], in 1956, reported that, although further work would be necessary before definitive conclusions may be drawn, their new results were most easily explained by assuming the existence in various minerals of a long-lived technetium isomer of mass 98.

3.7. Absence of Primordial Technetium in the Earth's Crust

In 1956, Boyd and Larson [3.22] at the Oak Ridge National Laboratory reported that a search for primordial technetium failed to reveal any traces of this element in a variety of terrestrial substances. They were unable to confirm the reports by previous investigators of the occurrence of technetium in molybdenite and yttrotantalite.

They stated that radioactive technetium almost certainly exists in certain terrestrial substances in extremely minute quantities: in uranium ores by virtue of the spontaneous fission of ^{238}U and by neutron-induced fission of ^{235}U; in molybdenum-containing minerals as a result of the capture of cosmic ray neutrons; and possibly in other substances, as the end product of extremely high energy reactions caused by other components of cosmic ray neutrons; and possibly in other substances, as the end product of extremely high energy reactions caused by other components of cosmic rays. However, they limited their work to the question of whether or not primordial technetium remains on the Earth today. Prior to 1956, the half-life of ^{98}Tc was only known to be *long*, but a value of 1.5×10^6 years was reported by O'Kelley and co-workers [3.23].

A vastly improved radiochemical procedure was used by Boyd and Larson. Table 3.2 shows the experimental results obtained. Apparently technetium was detected in but two

3.20   E.A. Alperovitch and J.M. Miller, Nature 176:299 (1955)
3.21   E. Anders, R.N. Sen Sarma, and P.H. Kato, J. Chem. Phys. 24:622 (1956)
3.22   G.E. Boyd and Q.V. Larson, J. Phys. Chem. 60:707 (1956)
3.23   G.D. O'Kelley, N.H. Lazar and E. Eichler, Phys. Rev. 101:1059 (1956)

Table 3.2. Search for technetium in various terrestrial minerals by Boyd and Larson (1956)

No.	Type of material	Wt. processed (g.)	% Tc recovered	Method of analysis for Tc	Technetium content of final concentrate (g.)	Technetium concentration in starting material	Rhenium concn. in starting material (p.p.m.)
16	MoS$_2$ concentrate	1000	25	Spectrochemical	$<10^{-7}$	$<4 \cdot 10^{-10}$	20
:	Pure KReO$_4$	30	~40	Spectrochemical	$<10^{-7}$	$<8 \cdot 10^{-9}$	:
:	Pure KReO$_4$	25	50	Spectrochemical	$<10^{-7}$	$<8 \cdot 10^{-9}$	:
4	MoS$_2$ concentrate	100	67.5	Spectrochemical	$<10^{-7}$	$<1.5 \cdot 10^{-9}$	1030
21	MoS$_2$ concentrate	100	27,5	Spectrophotometric	$<5 \cdot 10^{-6}$	$<1.8 \cdot 10^{-7}$	152
21	MoS$_2$ concentrate	250	No tracer	Spectrochemical	$<10^{-7}$	....	152
21	MoS$_2$ concentrate	250	55.9	Spectrochemical	$<10^{-7}$	$<7.6 \cdot 10^{-10}$	152
:	Osmiridium concentrate	10	63.4	Spectrochemical	$<10^{-7}$	$<1.6 \cdot 10^{-8}$	:
21	MoS$_2$ concentrate	100	74.9	Spectrochemical	$\sim6 \cdot 10^{-6}$	$6 \cdot 10^{-8}$	152
17	Flue dust	10	79.3	Spectrochemical	$<10^{-7}$	$<1.3 \cdot 10^{-8}$	3130
17	Flue dust	100	69	Spectrochemical	$<10^{-7}$	$<1.5 \cdot 10^{-9}$	3130
23	MoS$_2$ concentrate	184	No tracer	Activation	$<10^{-9}$	....	150
4	MoS$_2$ concentrate	100	76	Activation	$\sim10^{-8}$	$\sim1.3 \cdot 10^{-10}$	1030
22	MoS$_2$ concentrate	100	60.5	Spectrochemical	$<10^{-7}$	$<1.7 \cdot 10^{-9}$	297
:	Iron-nickel meteorite			Activation	$\sim5 \cdot 10^{-9}$	$\sim8.3 \cdot 10^{-11}$	:
		1000	22.4	Spectrochemical	$<10^{-7}$	$<4.5 \cdot 10^{-10}$	2
8	MoS$_2$ concentrate	1000	86	Mass spectrometric	$<4 \cdot 10^{-8}$	$<4.6 \cdot 10^{-11}$	688
4	MoS$_2$ concentrate	885	92	Activation	$<10^{-9}$	$<1.2 \cdot 10^{-12}$	1030
				Mass spectrometric	$<5 \cdot 10^{-8}$	$<6.1 \cdot 10^{-11}$	:
25	Yttrotantalite	100	68	Mass spectrometric	$<3.7 \cdot 10^{-7}$	$<5.4 \cdot 10^{-9}$	:

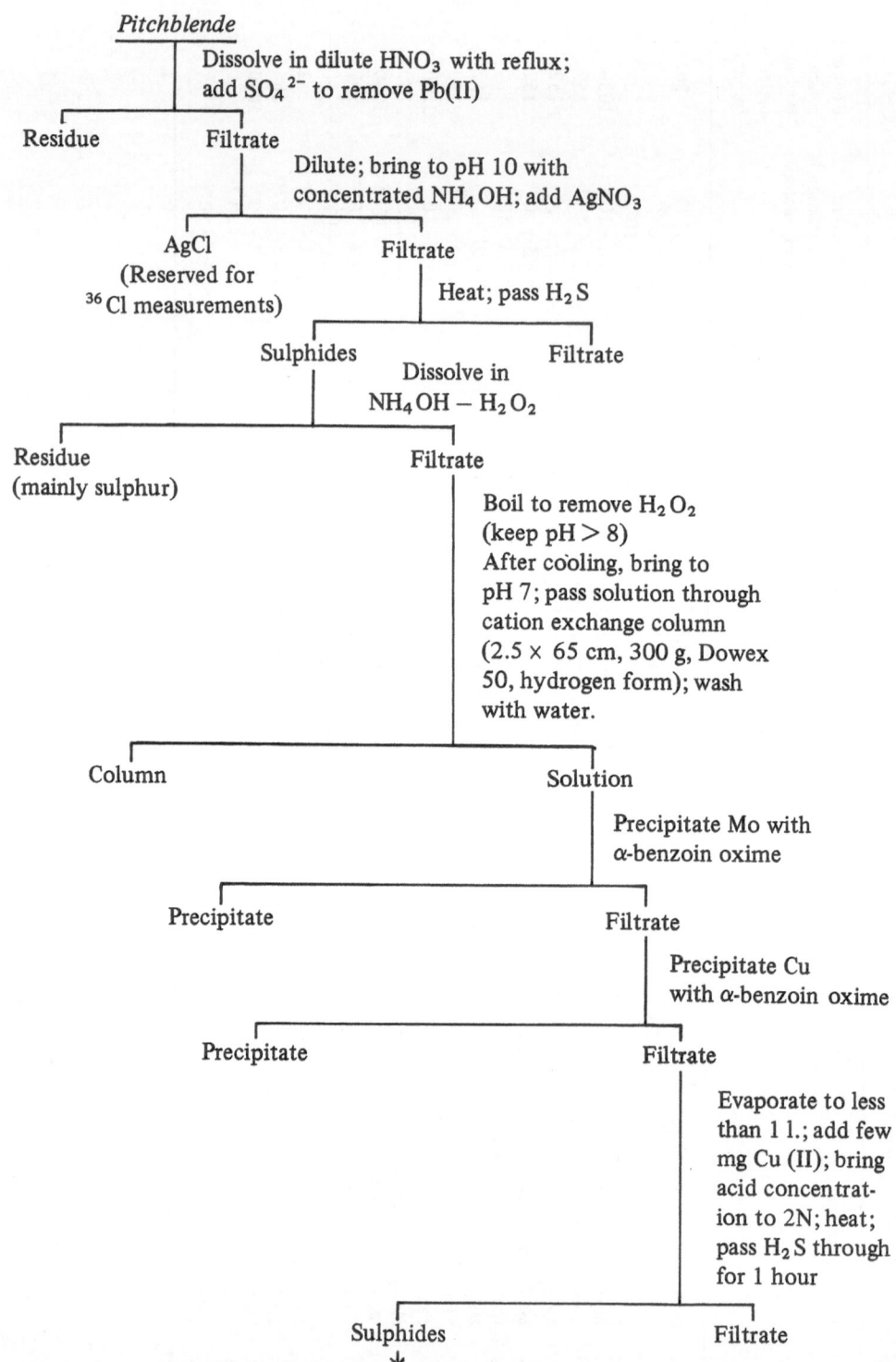

Fig. 3.2. Flow diagram for the isolation of Tc from pitchblende. (B.T. Kenna and P.K. Kuroda, J. Inorg. Nucl. Chem 26:493 (1964))

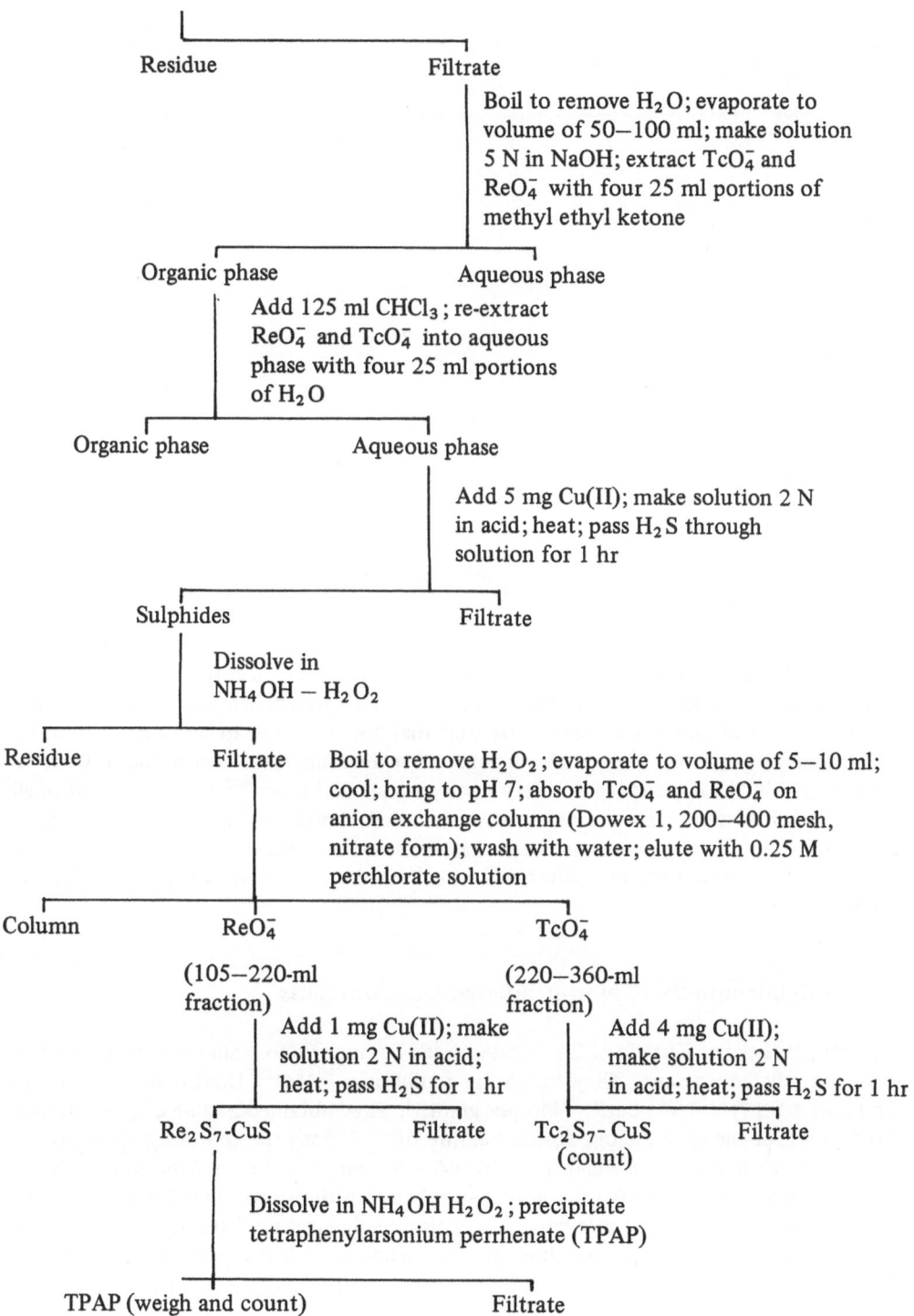

Residue   Filtrate

Boil to remove H_2O; evaporate to
volume of 50–100 ml; make solution
5 N in NaOH; extract TcO_4^- and
ReO_4^- with four 25 ml portions of
methyl ethyl ketone

Organic phase   Aqueous phase

Add 125 ml $CHCl_3$; re-extract
ReO_4^- and TcO_4^- into aqueous
phase with four 25 ml portions
of H_2O

Organic phase   Aqueous phase

Add 5 mg Cu(II); make solution 2 N
in acid; heat; pass H_2S through
solution for 1 hr

Sulphides   Filtrate

Dissolve in
$NH_4OH - H_2O_2$

Residue   Filtrate

Boil to remove H_2O_2; evaporate to volume of 5–10 ml;
cool; bring to pH 7; absorb TcO_4^- and ReO_4^- on
anion exchange column (Dowex 1, 200–400 mesh,
nitrate form); wash with water; elute with 0.25 M
perchlorate solution

Column ReO_4^- TcO_4^-

(105–220-ml   (220–360-ml
fraction)   fraction)

Add 1 mg Cu(II); make   Add 4 mg Cu(II);
solution 2 N in acid;   make solution 2 N
heat; pass H_2S for 1 hr   in acid; heat; pass H_2S for 1 hr

Re_2S_7-CuS   Filtrate Tc_2S_7-CuS   Filtrate
                                                 (count)

Dissolve in $NH_4OH \ H_2O_2$; precipitate
tetraphenylarsonium perrhenate (TPAP)

TPAP (weigh and count)   Filtrate

Fig. 3.2 continued

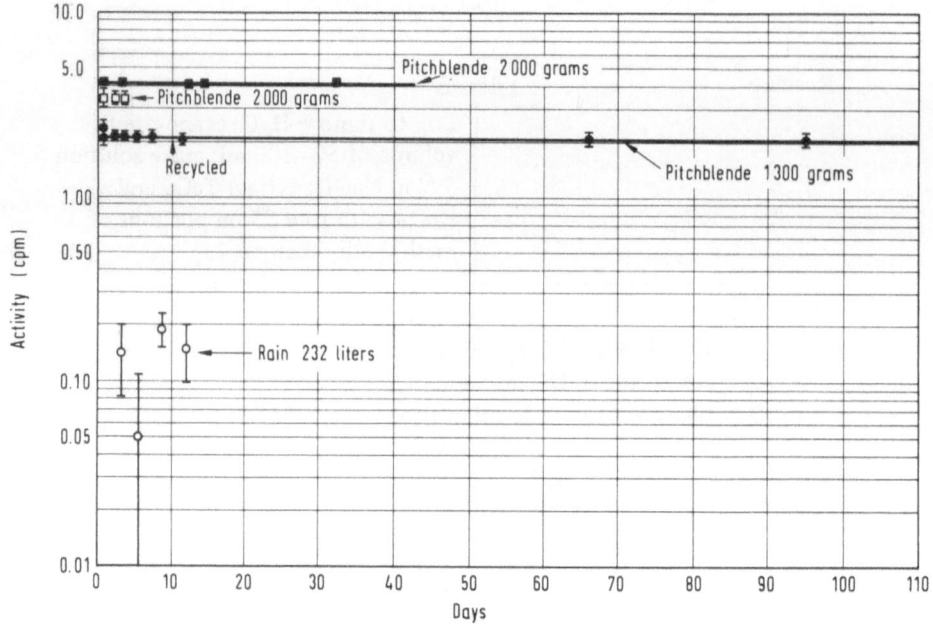

Fig. 3.3. Technetium activities isolated from pitchblende. (B.T. Kenna and P.K. Kuroda, J. Inorg. Nucl. Chem. 26:493 (1964))

instances. The most striking of these was with molybdenite concentrate No. 21 where as much as 6 micrograms of technetium were detected spectrochemically. Subsequently, however, neutron activation analysis showed that this was due to a contamination by [99]Tc. Minute quantities (0.005 micrograms) of technetium were also found in Concentrate No. 22 by the appearance of a very small amount of 6-hour ^{99m}Tc activity, but small quantities of rhenium were present so that this result could not be accepted without reserve. Boyd and Larson thus concluded that either technetium was absent, or, if present, it occurred in concentrations below those which could be detected by the means employed in their investigation.

3.8. Molybdenum-99 in Non-Irradiated Uranium Salts

In 1956, Parker and Kuroda [3.24, 3.25] isolated 67-hour [99]Mo from a kilogram quantity of non-irradiated uranium. They found the equilibrium [99]Mo/[238]U ratio in the sample to be $(1.26 \pm 0.12) \times 10^{-14}$ curie [99]Mo per gram U. They obtained a value of $(8.4 \pm 0.8) \times 10^{15}$ years for the spontaneous fission half-life of [238]U with the following assumptions: (a) the spontaneous fission yield of [99]Mo is 6.2 percent, (b) the contribution of the induced fission is small, and (c) the interchange between the fission molybdenum and the added molybdenum carrier was complete. This experiment paved the way for the isolation of ^{99}Tc from kilogram quantities of pitchblende described in the following section.

3.24   P.L. Parker and P.K. Kuroda, J. Chem. Phys. 25:1084 (1956)
3.25   P.L. Parker and P.K. Kuroda, J. Inorg. Nucl. Chem. 5:153 (1958)

3.9. Technetium in Pitchblende

Kenna and Kuroda [3.26, 3.27, 3.28] reported in 1961 and 1964, that a sample of African pitchblende was found to contain between 2.5 and 2.1×10^{-10} gram of ^{99}Tc per kilogram ore, but no other technetium isotopes were detected. The ^{99}Tc/^{238}U ratio in pitchblende was about the same as the ^{99}Mo/^{238}U ratio in non-irradiated uranium. The experimental results thus suggested that the ^{99}Tc is produced in the ore predominantly by ^{238}U spontaneous fission.

Three technetium fractions were isolated from kilogram quantities of the Belgian Congo pitchblende. The samples No. 1, No. 2 and No. 3 weighed 2, 1.3 and 2 kilograms, respectively. A flow sheet of the chemical operations is shown in Fig. 3.2.

The counting results of the technetium fractions isolated from pitchblende are shown in Fig. 3.3, together with the counting results of the technetium fraction isolated from 232 liters of rain obtained by Attrep [3.29]. Sample No. 1 was counted for about ten days and then recycled. The activity remained approximately the same after the recycling process, and no sign of decay was observed after being counted for about 100 days.

Sample No. 3 was counted for about a month and it also showed no sign of decay. The technetium fraction isolated from rain showed a barely detectable activity indicating that the possibility of contaminations from radioactive fallout was small during the course of this investigation.

An attempt was made to weigh the technetium fraction in Sample No. 2 to see if a weighable quantity of technetium had been isolated from pitchblende. Copper(II) sulphide was dissolved in a cyanide solution and removed from the technetium fraction. No weighable amount of precipitate remained in the technetium fraction and it was concluded that the technetium fraction isolated from pitchblende weighed less than 2 μg.

Table 3.3 summarizes the experimental results. An estimated error of about ± 25 percent, derived from the errors associated with chemical yield, counting efficiency, etc., is to be assigned to each value listed in this table.

Table 3.3. Tc^{99} in pitchblende, according to Kenna and Kuroda (1964)

Sample No.	Observed activity (counts/min per kg ore)	Tc^{99} content (10^{-12} Curies per kg ore)	Tc^{99} content (10^{-10} g Tc^{99} per kg ore)
1	1.8	4.4	2.5
2	1.7	4.3	2.5
3	2.1	5.4	3.1

3.26   B.T. Kenna and P.K. Kuroda, J. Inorg. Nucl. Chem. 23:142 (1961)
3.27   B.T. Kenna, J. Chem. Ed. 39:436 (1962)
3.28   B.T. Kenna and P.K. Kuroda, J. Inorg. Nucl. Chem. 26:493 (1964)
3.29   M. Attrep, Jr., M.S. Dissertation, University of Arkansas (1962)

Table 3.4. Mo^{99} in non-irradiated uranium and Tc^{99} in pitchblende

Nuclide	Uranium Sample	10^{-4} (Disintegrations) per sec per g uranium)
Mo^{99}	In non-irradiated natural uranium	4.4 ± 0.4
Mo^{99}	In non-irradiated depleted uranium	4.4 ± 0.3
Tc^{99}	In pitchblende	$4 \quad \pm 1$

The $^{99}Tc/^{238}U$ ratio found in pitchblende agrees fairly well with the $^{99}Mo/^{238}U$ ratios in non-irradiated natural and depleted uranium salts measured by Parker and Kuroda [3.24, 3.25], as shown in Table 3.4.

These results indicate that the ^{99}Tc is produced in pitchblende predominantly by the spontaneous fission of ^{238}U. The activity of ^{99}Tc which is expected to be present in uranium ore is

$$_{99}A = {_{99}Y} \cdot {^{238}N} \cdot {_{238\,f}\lambda} \tag{3.3}$$

where $_{99}A$ is the activity of ^{99}Tc, $_{99}Y$ is the fission yield for the mass 99 chain, ^{238}N is the number of atoms of ^{238}U, and $_{238\,f}\lambda$ is the spontaneous fission decay constant of ^{238}U. Equation (3.3) yields a value of 4.8×10^{-12} curies ^{99}Tc per kilogram ore, which is in agreement with the experimental data shown in Table 3.4.

3.10. The Occurrence of Promethium Isotopes in Non-Irradiated Uranium Salts

In 1964, Menon and Kuroda [3.30] reported that they were able to isolate ^{147}Pm and ^{149}Pm from kilogram quantities of non-irradiated uranium salts. They adopted the ion exchange method of separation of rare earth nuclides reported by Choppin and co-workers [3.31, 3.32]. About 2 kilograms of non-irradiated uranyl nitrate were dissolved in ether, and the solution was shaken with 4 M nitric acid containing cerium carrier. Uranium and thorium were removed by TBP (tri-n-butyl phosphate) extraction. The carrier cerium and other lanthanoid activities were then precipitated as the hydroxide. The promethium fraction was then separated by the ion exchange separation method.

The equilibrium ratios of the fission products/^{238}U in non-irradiated uranium are shown in Table 3.5. These ratios were in close agreement with the value calculated from the spontaneous fission half-life of ^{238}U.

3.30   M.P. Menon and P.K. Kuroda, J. Inorg. Nucl. Chem. 26:401 (1964)
3.31   G.R. Choppin, B.G. Harvey and S.G. Thompson, J. Inorg. Nucl. Chem. 2:66 (1956)
3.32   G.R. Choppin and R.J. Silva, J. Inorg. Nucl. Chem. 3:153 (1956)
3.33   M. Attrep, Jr., and P.K. Kuroda, J. Inorg. Nucl. Chem. 30:699 (1968)
3.34   O. Erämetsä, Acta Polytechnica Scad. Chem. 37:1 (1965)

Table 3.5. Equilibrium ratios of the fission products/^{238}U in non-irradiated uranium reported by Menon and Kuroda in 1964

Mass number	Nuclide measured	Number of measurements	Equilibrium ratio (10^{-4} disintegration sec^{-1} g^{-1} U)
141	Ce	2	4.9 ± 1.0
143	Ce	4	4.4 ± 0.8
144	Ce	2	4 ± 1
147	Nd	2	3.1 ± 0.8
147	Pm	2	3.6 ± 0.9
149	Pm	3	1.6 ± 0.2

Table 3.6. Promethium in pitchblende, according to Attrep and Kuroda (1968)

Sample	Promethium-147 content	
	(10^{-4} disintegrations/ sec/g U)	(10^{-15} g/kg ore*)
Pitchblende No. 1 (1200 g)	3.5 ± 1.0	4.4 ± 1.3
Pitchblende No. 2 (1500 g)	3.7 ± 0.7	4.6 ± 0.9
Natural uranium salt+	3.6 ± 0.9	—

* U content: 42.1 per cent
+ Data obtained by Menon and Kuroda (1964)

3.11. Promethium in Pitchblende

Attrep, and Kuroda [3.33] reported in 1968, that a sample of African pitchblende was found to contain (4 ± 1) × 10^{-15} grams of ^{147}Pm per kilogram of ore. The observed ^{147}Pm/U ratio in the pitchblende was in agreement with the ^{147}Pm/U equilibrium ratio in non-irradiated natural uranium as shown in Table 3.6. Their results thus indicated that the ^{147}Pm in African pitchblende is produced predominantly by ^{238}U spontaneous fission.

It is interesting to note here that Erämetsä [3.34] reported that he isolated ^{147}Pm from a rare earth oxide concentrate representing 6,000 tons of apatite. The activity isolated was 3.1 × 10^2 disintegrations per second. The origin of this ^{147}Pm was attributed to neodymium bombardment by cosmic-ray neutrons.

A brief outline of the investigations on the natural occurrences of the elements 43 and 61 has been presented in this chapter. The results from these investigations seemed to indicate that the terrestrial abundances of these elements are essentially controlled by their production by the spontaneous fission of ^{238}U. In other words, these elements occur in the earth's crust in the same sense that, for example, radium and polonium occur in the lithosphere in radioactive equilibria with their long-lived precursor ^{238}U. The story does not end here, however, as we shall see in the next chapter that elements 43 and 61 once existed on the earth in far greater quantities than the amounts found in the uranium ore deposits today (see Chapter 4: The Oklo Phenomenon).

4.
The Oklo Phenomenon

"In this event the whole of the hydrogen on the earth might be transformed at once and the success of the experiment published at large to the Universe as a new star."

Francis W. Aston, December 12, 1922

Until recently, scientists believed that the chemical elements were synthesized only in stars. The discovery of the Oklo phenomenon in 1972 has revealed, however, that a nuclear "fire" had existed on the earth and large-scale transmutations of the elements were occurring on our planet billions of years ago.

4.1. Discovery of Spontaneous Fission

In 1939, Libby [4.1] attempted to discover the process of spontaneous fission by performing the following experiment: to a slightly acidified aqueous solution of 0.90 mole of uranyl nitrate which had not been disturbed chemically for at least 5 years, a little solid iodine was added and was extracted, shaking with CCl_4. The iodine was precipitated as AgI in a thin layer and was counted in a screen wall counter so no beta-radiation harder than 20,000 electron volts would have been missed. The iodine sample thus obtained showed a radioactivity of 0 ± 20 counts per minute.

If it was assumed that each ten fission events produced at least one radioactive iodine on the average, this result required that the half-life for natural fission be at least as long as 10^{14} years. A similar experiment was performed on thorium with the same results.

In order to investigate the possibility of natural neutron emission from uranium and thorium, a BF_3-filled counter surrounded with paraffin and standardized with 200 mg of $RaBr_2$ mixed with Be powder was used. Seven and one-half moles of uranium salts were

---

4. 1 W.F. Libby, Phys. Rev. 55:1269 (1939)

found to give less than two counts per minute, whereas the Ra-Be source in the same position gave 4600 counts per minute.

The discovery of spontaneous fission of uranium was announced by the Soviet investigators Flerov and Petrzhak [4.2]. The manuscript was sent to *The Physical Review* by cable, and the entire text reads as follows:

"Spontaneous fission of Uranium. With 15 plates ionization chambers adjusted for detection of uranium fission products we observed 6 pulses per hour which we ascribe to spontaneous fission of uranium. A series of control experiments seem to exclude other possible explanations. Energy of pulses and absorption properties coincide with fission products of uranium bombarded by neutrons. No pulses were found with UX and Th. Mean lifetime of uranium follows ten to sixteen or seventeen years. Flerov, Petrzhak, Physico Technical Institute (F), Radium Institute (P), Leningrad, U.S.S.R., June 14, 1940.

Although Libby [4.1] missed the discovery of spontaneous fission, he put forward an intriguing idea in 1939. He stated: "Natural fission might be detected by disparities between the results of various calculations of the age of the oldest rocks based on the observed abundances of the elements and the intensities of the ordinary radioactive emissions from uranium and thorium. The extremely long half-life observed for the natural fission of uranium, however, seemed to exclude such a possibility. The agreement between the uranium and thorium age determinations on old rocks and the agreement of these with other estimates of the age of the oldest rocks seem to indicate that the age of the earth is at least 2×10^9 years and no fission process with a half-life less than 5×10^9 years occurs, *unless uranium and thorium have been or are now being generated by some terrestrial process.*"

4.2. Plutonium-239 in Nature

During the period between 1941 and the early part of 1942, Seaborg and Perlman [4.3] investigated the occurrence of elements 93 and 94 in nature. In a paper which was written as a secret report and mailed to the "Uranium Committee" in Washington, D. C., on April 13, 1942, and published six years later in *The Journal of American Chemical Society*, they stated: "The hope would be to discover a very long-lived 94 or 93 and that this be present in an amount large enough so that useful quantities could be extracted from the minerals."

A 400-gram sample of pitchblende concentrate from the Great Bear Lake region of Canada was used in this investigation. A chemical method for separating and concentrating elements 93 and 94 from uranium and thorium was applied to the sample of pitchblende. A final fraction of 93 and 94 precipitated with rare earth carrier was counted for fissions with slow and with fast neutrons, and an upper limit of 1 part in 10^8 to 10^9 has been set for the amount of these elements in the sample. Based on the number of alpha-particle counts in the sample, they estimated the ^{239}Pu content to be 1 part in 10^{14} in the original pitchblende concentrate. This amounted to only a few percent of the amount to

4. 1   W.F. Libby, Phys. Rev. 55:1269 (1939)
4. 2   Flerov and Petrzhak, Phys. Rev. 58:89 (1940); K.A. Petrzhak and G.N. Flerov, Compt. Rend. Akad. Sci. USSR 25:500 (1940)
4. 3   G.T. Seaborg and M.L. Perlman, J. Am. Chem. Soc. 70:1571 (1948)

be expected if a large proportion of the spontaneous fission neutrons were absorbed by the ^{238}U to produce ^{239}Pu:

$$^{238}U + \text{spontaneous fission neutron} \rightarrow {}^{239}U$$

$$\overset{\beta^-}{\rightarrow} \quad {}^{239}93 \quad \overset{\beta^-}{\rightarrow} \quad {}^{239}Pu.$$

It appeared to the writer that the natural occurrence of ^{239}Pu in uranium ores was proof *that sub-critical uranium chain reactions are actually occurring in nature*. Would it not be possible then that the "disparities" such as envisioned by Libby might still be observed, if self-sustaining uranium chain reactions had occurred in nature during the geological history of the earth? Would it not be also possible that the terrestrial abundances of elements 43 and 61 might have been significantly greater than the values calculated from the spontaneous fission half-life of ^{238}U, if natural reactors had existed on the earth?

4.3. Large-Scale Nuclear Processes on the Earth

The possibility of the occurrence of large-scale nuclear reactions in the earth's crust or mantle was discussed by J. Noetzlin [4.4] in 1939, by M. Odagiri [4.5] in 1940, and by N. Efremov [4.6] in 1946. J.B. Orr [4.7] was probably the first to consider the possibility of the existence of nuclear reactors in nature. In 1949 and 1950, he investigated the ^{235}U content of thucholite from the Besner mine, Ontario, Canada, for vestiges of natural nuclear chain reactions. The results of his experiment, however, did not validate his hypothesis that a chain reaction might have taken place in the Canadian thucholite if its mass had been sufficiently great, and the carbon contained in the mineral had acted as a neutron moderator.

In 1951, R.R. Edwards at the University of Arkansas was initiating a research project to investigate the contents of radioactive fission products in the water of Hot Springs National Park, Arkansas. His hypothesis was that a uranium chain reaction might be occurring in the earth crust in the Hot Springs region. Initial attempts to prove the hypothesis of Edwards by one of his students, J.H. Jonte [4.8] and the writer during the early 1950's by measuring the contents of ^{89}Sr and ^{90}Sr in thermal waters of Hot Springs, Arkansas, led to a negative result.

4. 4   J. Noetzlin, Comptes rendus 208:1662 (1939); J. Phys. radium 1:8, 90 (1940); ibid.:124 (1940)
4. 5   M. Odagiri, On the parallel relationship between the developments of atoms and the earth as a heavenly body, Solar Industry Co. (in Japanese) 1940, pp. 1–52; J. Chem. Soc. Japan 63:224, 1425 (1942)
4. 6   N. Efremov, Die Entwicklung der Chemischen Elemente, "UNRRA" – University, Students' Union, München: 1–80 (1946)
4. 7   J.B. Orr, Phys. Rev. 76:155 (1949); 79:401 (1950)
4. 8   J.H. Jonte, PhD Dissertation, The University of Arkansas, 1956

4.4. Xenon Isotopes in Radioactive Minerals

In 1947, the Soviet investigators Khlopin and co-workers [4.9] reported that spontaneous fission of ^{238}U produces xenon in uranium minerals. Assuming that the products of spontaneous fission of uranium were identical to those produced under the action of slow neutrons and that 14% of the fission resulted in formation of stable xenon isotopes, they reported that it was possible to determine the age of the minerals from the xenon-uranium ratio.

In 1948, Khlopin and Gerling [4.10] calculated the amount of xenon that could have accumulated in the earth's crust during its existence. They found that the amount of xenon formed by fission amounted to about 0.5% of the total xenon content in the earth's crust, including the atmosphere, or several tens of millions of cubic meters of fission xenon.

During the early 1950's, however, mass-spectrometric studies on the krypton and xenon isotopes in various radioactive minerals revealed that neutron-induced fission of ^{235}U is also a source of xenon. In these studies, carried out by Thode and co-workers [4.11] and Wetherill [4.12], the mass-yield distribution of fission xenon extracted from the minerals was shown to be a mixture from spontaneous fission of ^{238}U and neutron-induced fission of ^{235}U, the contribution from the latter decreasing as the rare-earth content of the minerals increased.

In 1953, Fleming and Thode [4.11] measured the relative abundances of the stable isotopes of xenon and krypton in six samples of pitchblende and one sample of uranite and found that the fission yields observed varied markedly from sample to sample. In general the xenon fission yields most closely resembling those for neutron-induced fission of ^{235}U were associated with either very high uranium content (Belgian Congo pitchblende) or with a combination of high uranium content and old geological age (Eagle Mine pitchblende from Canada). The smallest contribution from the neutron-induced fission of ^{235}U was observed in a sample of rare earth-rich uraninite from Cardiff Township, Ontario, Canada. It thus appeared that these trends could be accounted for by the large neutron capture cross sections of some of the rare earth elements.

In 1953, Wetherill [4.12] reported on the isotopic compositions of xenon and krypton extracted from a sample of rare earth-rich mineral euxenite from Madagascar. The sample contained abount one percent each of gadolinium and dysprosium, and Wetherill estimated that only about one in three thousand of the neutrons captured in the mineral would be captured in ^{235}U. This meant that the contribution from the neutron-induced fission of ^{235}U must be negligibly small and hence the relative abundances of the xenon and krypton isotopes should represent the yields from the spontaneous fission of ^{238}U. Table 4.1 shows the ^{238}U spontaneous fission yield thus obtained from the isotopic compositions of xenon and krypton found in euxenite.

Wetherill also studied the xenon and krypton extracted from a 20-gram sample of Belgian Congo pitchblende containing 44 percent uranium. By using the values of ^{238}U

4. 9   V.G. Khlopin, E.K. Gerling and N.V. Baranovskaya, Bull. Acad. Sci. U.S.S.R. Classe Sci. chim. 1947:599; C.A. 42:3664
4.10   V.G. Khlopin and E.K. Gerling, Doklady Akad. Nauk (U.S.S.R.) 61:297 (1948); C.A. 43:527
4.11   W.H. Fleming and H.G. Thode, Phys. Rev. 92:378 (1953); see also, J. MacNamara and H.G. Thode, Phys. Rev. 80:471 (1950); and B.G.Young and H.G. Thode, Can. J. Phys. 38:1 (1960)
4.12   G.W. Wetherill, Phys. Rev. 92:907 (1953)

Table 4.1. Uranium-238 fission yields for xenon and krypton isotopes reported by Wetherill in 1953

Mass Number	Yields (percent)
83	0.036 ± 0.015
84	0.119 ± 0.040
86	$0.75 \ \pm 0.11$
129	< 0.012
131	0.455 ± 0.02
132	$3.57 \ \pm 0.06$
134	$4.99 \ \pm 0.07$
136	6.00 (assumed)

spontaneous fission yields as determined from the euxenite sample and the ^{235}U slow-neutron fission yields reported by Thode and Graham [4.13], he calculated the relative contributions of neutron-induced and spontaneous fission in Belgian Congo pitchblende. The result indicated that about 35 percent of the fission was neutron-induced. It is interesting to note that Wetherill and Inghram [4.14], in 1953, stated: "Thus the deposit was twenty-five percent of the way to becoming a pile. It is also interesting to extrapolate back 2,000 million years where the ^{235}U abundance was 6 percent instead of 0.7. Certainly, such a deposit would be closer to being an operating pile."

4.5. Radioactive Strontium Isotopes in Pitchblende

In 1954, Kuroda and Edwards [4.15] isolated 40 mg of strontium (as the carbonate) without added carrier from 2.5 kilograms of Great Bear Lake pitchblende (41.70 percent U_3O_8). The alkaline earths content was first determined by the analytical procedures described in "Applied Inorganic Analysis" by Hillebrand, Lundell, Bright and Hoffman [4.16], using a few gram sample of the pitchblende. The alkaline earths were then isolated from a 2.5 kilogram sample of pitchblende by performing the same chemical operations on a larger scale and repeating the experiments many times. Strontium was then separated from calcium and barium using the procedures described in the above-mentioned book.

The strontium sample thus obtained was further decontaminated from radium and barium activities by repeated additions of barium carrier, followed by barium chromate precipitations, until the barium chromate activity became negligible. Similar operations were performed with bismuth carrier (precipitated as the sulfide), and lanthanum carrier (precipitated as the hydroxide) until all activity following these carriers was removed. The strontium was then precipitated as the carbonate and mounted on aluminium for measurement with an end window Geiger tube. Recovery of the strontium was about 65 percent.

4.13   H.G. Thode and R.L. Graham, Can. J. Research A 25:1 (1947)

4.14   G.W. Wetherill and M.G. Inghram, Nuclear Processes in Geological Settings, The University of Chicago, 1953, p. 30

4.15   P.K. Kuroda and R.R. Edwards, J. Chem. Phys. 22:1940 (1954)

4.16   W.F. Hillebrand, G.E.F. Lundell, H.A. Bright and J.I. Hoffman, Applied Inorganic Analysis, 2nd Edition, John Wiley and Sons, Inc., New York

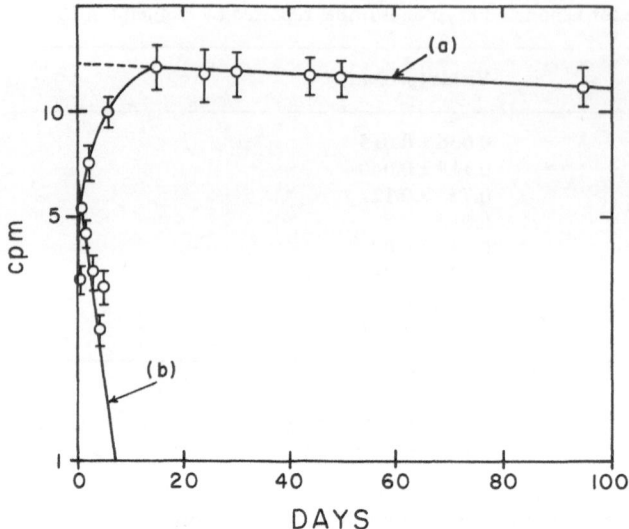

Fig. 4.1. Growth and Decay of Sr fraction isolated without added carrier from 2.5 kilogram of Great Bear Lake pitchblende by P.K. Kuroda and R.R. Edwards, *J. Chem. Phys. 22,* 1940 (1954) (a) Growth of ^{90}Y, decay of ^{89}Sr; (b) decay of ^{90}Y

Results of the radioactivity measurements are shown in Fig. 4.1. The initial rise is due to the growth of the 61-hour ^{90}Y, and the final decay appears to be that of the 55-day ^{89}Sr. The sample was re-dissolved, and a lanthanum hydroxide precipitation was performed, yielding a sample showing, within experimental error, the decay of ^{90}Y.

The amont of ^{90}Sr found in the pitchblende was $(1.3 \pm 0.1) \times 10^{-14}$ curie per gram of ^{238}U. This corresponded to an overall fission half-life of $(5.9 \pm 0.6) \times 10^{15}$ years, if a fission yield for ^{90}Sr of 5 percent was assumed. The radiochemically determined fission half-life appeared to be some 25 percent shorter than the value of $(8.04 \pm 0.3) \times 10^{15}$ years for the spontaneous fission of ^{238}U reported by Segrè [4.17]. Kuroda and Edwards interpreted this to be due to the contribution of neutron-induced fission, and noted that this compared with the value of 35 percent contribution from the neutron-induced fission reported by Wetherill [4.12] for Belgian Congo pitchblende.

Similar experiments were performed by Kuroda and Edwards [4.18], in which extremely low ^{140}Ba activities were isolated from 4,540 grams of non-irradiated uranyl nitrate. The equilibrium amount of ^{140}Ba in the uranium salt was found to be $(1.6 \pm 0.1) \times 10^{-14}$ curie per gram ^{238}U. The experiment was repeated by Heydegger and Kuroda [4.19] in 1959 with both natural and depleted uranium.

4.17   E. Segrè, Phys. Rev. 86:21 (1952)

4.18   P.K. Kuroda and R.R. Edwards, J. Inorg. Nucl. Chem. 3:345 (1957)

4.19   H.R. Heydegger and P.K. Kuroda, J. Inorg. Nucl. Chem. 12:12 (1959)

4.6. Iodine-129 in Pitchblende

In 1956, Purkayastha and Martin [4.20] reported on the results from their studies on the natural production of ^{129}I in pitchblende. They used the activation analysis method for the determination of ^{129}I and compared the results with those for ^{129}Xe determined mass-spectrometrically. Expressing the results in terms of

$$\frac{\text{disintegrations of } ^{129}\text{I per second}}{\text{spontaneous fission per second}},$$

they reported the ^{129}I yields to be 0.48 (sample A) and 0.54 (sample B) percent. These values were much greater than the ^{129}Xe yields obtained mass-spectrometrically: 0.222 (sample A) and 0.301 (sample B) percent. Although the reason for the discrepancy could not be given at that time, it appeared as if the data indicated that the contribution from the neutron-induced fission of ^{235}U in the pitchblende was quite appreciable.

4.7. Radioactive Iodine Isotopes in Aqueous Uranium Solutions

The experiment to isolate extremely small activities of radioactive iodine isotopes from large amounts of non-irradiated uranium salts, which was first performed by Libby [4.1] in 1939, was repeated in the writer's laboratories in 1956 [4.21, 4.22]. Kilogram quantities of both natural and depleted (0.011 percent ^{235}U) uranium salts were used in the 1956 experiments. Radioactive iodine isotopes were periodically extracted from the aqueous solutions of non-irradiated uranium. By repeating many experiments, in which the time interval between two successive iodine extractions was varied from 30 minutes to 15 days, it was possible to vary the relative activities of the short-lived iodine isotopes obtained in the final iodine fractions:

$$^{131}\text{I} \xrightarrow{8 \text{ days}} {}^{131}\text{Xe (stable)}$$

$$^{132}\text{I} \xrightarrow{2.2 \text{ hours}} {}^{132}\text{Xe (stable)}$$

$$^{133}\text{I} \xrightarrow{21 \text{ hours}} {}^{133}\text{Xe} \xrightarrow{5.3 \text{ days}}$$

$$^{134}\text{I} \xrightarrow{52 \text{ minutes}} {}^{134}\text{Xe (stable)}$$

$$^{135}\text{I} \xrightarrow{6.7 \text{ hours}} {}^{135}\text{Xe} \xrightarrow{9.2 \text{ hours}}$$

It was felt at that time that some difference should be noticeable between the iodine activities produced from the aqueous solutions of natural and depleted uranium, because the contribution from the neutron-induced fission of ^{235}U in natural uranium is expected to be many times that in depleted uranium. The uranium salts were dissolved in several

4.20   B.C. Purkayastha and G.W. Martin, Can. J. Chem. 34:293 (1956)
4.21   P.K. Kuroda, R.R. Edwards and F.T. Ashizawa, J. Chem. Phys. 25:603 (1956)
4.22   F.T. Ashizawa and P.K. Kuroda, J. Inorg. Nucl. Chem. 5:12 (1957)

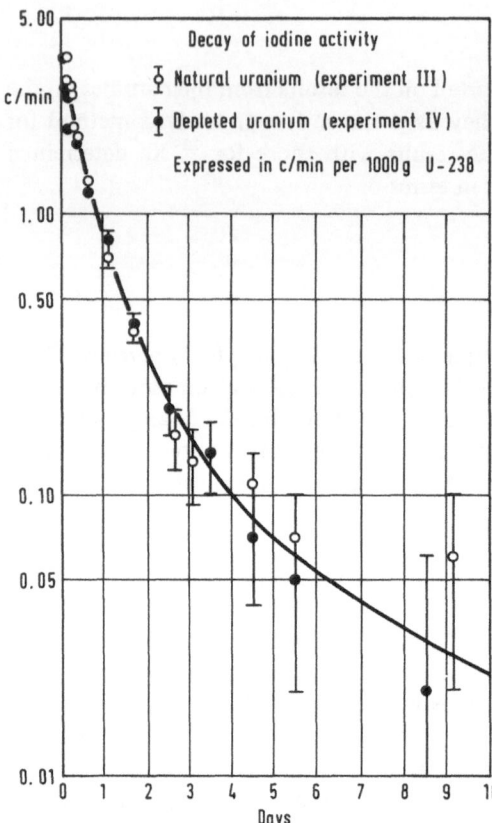

Fig. 4.2. The iodine isotopes/^{238} U equilibrium ratios in natural and in depleted uranium solution (Ashizawa and Kuroda, 1957). The data are expressed in counts per minute of short-lived iodine activities per 1,000 grams of ^{238} U in the solution

liters of water and the hydrogen atoms in water should have acted as an efficient moderator of the neutrons. No significant difference was observed, however, between the activities of the short-lived iodine isotopes produced in natural and depleted uranium, as shown in Fig. 4.2.

It appeared that the result obtained by Ashizawa and Kuroda [4.22] could be explained by the application of the nuclear reactor theory to the system of the aqueous solutions of natural and depleted uranium. The infinite multiplication constant (k_∞) is usually written in the form [4.23, 4.24]

$$k_\infty = \epsilon p f \eta = \epsilon \exp\left[-\frac{N_0}{\xi \Sigma_s} \int_E^{E_0} (\sigma_a)_{eff}\, dE/E\right] \times \qquad (4.1)$$

$$\frac{N_u (\sigma_a)_u}{N_u (\sigma_a)_u + N_m (\sigma_a)_m + N_1 (\sigma_a)_1 + N_2 (\sigma_a)_2 + \ldots} \times$$

$$\frac{^{235}N (\sigma_f)_{235}}{^{235}N (\sigma_a)_{235} + {}^{238}N (\sigma_a)_{238}} \cdot \nu$$

4.22   F.T. Ashizawa and P.K. Kuroda, J. Inorg. Nucl. Chem. 5:12 (1957)
4.23   R. Stevenson, Introduction to Nuclear Engineering, McGraw-Hill, New York, 1954
4.24   R.L. Murray, Introduction to Nuclear Engineering, Prentice-Hall, 1954

where ϵ is the fast fission factor, p is the resonance escape probability, f is the thermal utilization factor, η is the average number of fission neutrons produced per neutron absorbed by uranium, N_0 is the number of atoms of ^{238}U per cm^3, ξ is the average loss in logarithm of the neutron energy from an elastic collision, Σ_s is the macroscopic scattering cross-section of the moderator, $_E\int^{E_0} (\sigma_a)_{eff}\, dE/E$ is the effective resonance integral, N_u is the number of atoms of uranium per cm^3, N_m is the number of atoms of moderator (water) per cm^3, N_1, N_2 ... are the number of atoms of impurities per cm^3, (σ_a) is the microscopic thermal neutron absorption cross-section for material in question, ^{235}N is the number of atoms of ^{235}U per cm^3, ^{238}N is the number of atoms of ^{238}U per cm^3, $(\sigma_f)_{235}$ is the microscopic fission cross-section of ^{235}U, $(\sigma_a)_{235}$ is the microscopic absorption cross section of ^{235}U, and ν is the average number of fast neutrons emitted per thermal fission of ^{235}U.

Neglecting the effect of added reagents and the impurities, one obtains from the above equation, a value of $k_\infty = 0.13$ for an aqueous solution of natural uranium (1400 g uranium in 10 liters of 1 N nitric acid), and a value of $k_\infty = 0.003$ for an aqueous solution of depleted uranium (1600 g ^{238}U in 8 liters of 1 N nitric acid).

The effective multiplication constant is

$$k_{eff} = k_\infty \cdot L_f \cdot L_t = k_\infty \cdot e^{-k^2 \tau} \cdot (1 + k^2 L^2)^{-1} \tag{4.2}$$

where, L_f is the fraction of fast neutrons which do not escape before becoming thermal, L_t is the fraction of thermal neutrons which do not get out before absorption, k^2 is the size-shape factor or "buckling", τ is the age or square of the fast diffusion length, and L is the thermal diffusion length. Simple calculations yield the following values for an aqueous solution of uranium placed in an approximately cylindrical container of radius R = 15 cm and height H = 14 cm: $L_f = 0.084$ and $L_t = 0.62$. Hence, a value of $k_{eff} = 7 \cdot 10^{-3}$ is obtained for the natural uranium solution and a value of $k_{eff} = 1.6 \cdot 10^{-4}$ for the depleted uranium solution.

Ashizawa and Kuroda [4.22] thus concluded that it appeared quite reasonable that no significant difference was observed in their experiments between the fission iodine/^{238}U ratios in natural uranium and in depleted uranium.

4.8. Resonance Capture of Neutrons in Pitchblende

In 1956, Burkhardt [4.25] published a short article in *Die Naturwissenschaften*, in which he discussed the resonance capture reactions of neutrons occurring in uranium ore deposits. He considered the following chemical composition of a sample of Katanga pitchblende, which is found in Gmelins Handbuch der anorganischen Chemie [4.26]:

UO_2	UO_3	PbO	Fe_2O_3	MoO_3	Se	H_2O	Insoluble
30.60	57.00	6.39	0.67	0.32	0.40	4.10	0.17 (Weight %)

Assuming that (a) the ore deposit is sufficiently large so that the fraction of the neutrons which escape from the system is negligibly small and (b) the oxygen atoms are sole-

4.25 W. Burkhardt, Naturwissenschaften 42:534 (1956); Ann. Physik 20:184 (1957)
4.26 Gmelins Handbuch der anorganischen Chemie, 8. Aufl., System Nummer 55, Uran und Isotope, Verlag Chemie, Berlin, 1936, p. 14

ly responsible for the slowing down of the neutrons, the value of the resonance escape probability p can be calculated by using the semi-empirical formula for the homogeneous reactors such as found in Glasstone's "Principles of Nuclear Reactor Engineering" [4.27]. If such a calculation is carried out for the Katanga pitchblende, a value of $p = 0.0003$ is obtained. If, however, the hydrogen atoms, which are always present in pitchblende are considered, a more reasonable value of p is obtained. For example, in the above pitchblende, $N_H/N_U = 1.2$ and a value of $p = 0.56$ is obtained for the resonance escape probability. Such values of p explain the occurrence of ^{239}Pu in pitchblende and also the fact that the contribution of the neutron-induced fission of ^{235}U amounts to as high as 30 percent in the production of the fission xenon in some of the uranium ore deposits.

4.9. The Theory of Natural Reactors

The idea of naturally-occurring self-sustaining nuclear chain reactions became quite unpopular during the 1950's and remained so for many years. One of the reasons was that it appeared as if Fermi's pile theory [4.28], when applied to uranium ore deposits, seemed to lead to an unequivocal conclusion that the chain reaction could never have become self-sustaining. The fact that it is not necessarily so was pointed out in a brief article [4.29] entitled "On the Nuclear Physical Stability of the Uranium Minerals", which appeared in the October 1956 issue of *The Journal of Chemical Physics*. In this paper, the writer stated: "The infinite multiplication constant, k_∞, may be considered as an indicator of the stability of the uranium minerals, which are the natural assemblages of uranium, moderator, and impurities. We may consider a system to be quite 'stable', when the infinite multiplication constant of the assemblages is far less than unity. The system will be nuclear physically 'unstable', when the infinite multiplication constant is greater than unity."

According to Fermi's pile theory [4.28],

$$k_\infty = \epsilon\, p\, f\, \eta \tag{4.3}$$

where ϵ is the fast fission factor, p is the resonance escape probability, f is the thermal utilization factor, and η is the number of fast neutrons available per neutron absorbed by uranium. When dealing with geological events, the change of the uranium enrichment as a function of geological time should also be taken into consideration. The major neutron sources in minerals are the spontaneous fission of ^{238}U and the (α, n) reactions.

The values of p and f can be calculated if the chemical composition of the mineral is given, ϵ is always close to unity, and η as a function of the uranium enrichment is known. Hence the value of k_∞ of a mineral at any geological time can be calculated.

A pitchblende ore from Johanngeorgenstadt, Saxony, was chosen as an example by the writer in 1956. The reason for choosing this pitchblende was two-fold: (a) it appeared to be free of the neutron-consuming rare earth elements, which meant that this ore was likely to be the one, which would have most easily become nuclear physically "unstable"

4.27   S. Glasstone, Principles of Nuclear Reactor Engineering, Van Nostrand, New York, 1955
4.28   E. Fermi, Science 105:27 (1947)
4.29   P.K. Kuroda, J. Chem. Phys. 25:781 (1956)

Table 4.2. Analysis of Johanngeorgenstadt, pitchblende*

UO_3	22.33
UO_2	59.30
ThO_2	None
CeO_2	None
ZrO_2	None
$(La, Di)_2 O_3$	None
$(Yt, Er)_2 O_3$	None
$Al_2 O_3$	0.20
$Fe_2 O_3$	0.21
PbO	6.39
CuO	0.17
MnO	0.09
CaO	1.00
MgO	0.17
$Bi_2 O_3$	0.75
$V_2 O_5$, MoO_3, WO_3	0.75
Alkalies	0.31
SO_3	0.19
$P_2 O_5$	0.06.
$As_2 O_5$	2.34
He	Trace
$H_2 O$	3.17
SiO_2	0.50
Total	97.93

* Analyst: W.F. Hillebrand (Bull. U.S. Geol. Survey 78:43 (1891), 90:23 (1892); 220:111−114 (1903). See also, F.W. Clarke, The Data of Geochemistry, Washington Government Printing Office, 1924, p. 725

during the geological history of the earth; and (b) the chemical analysis was performed in the 1880's by W.F. Hillebrand [4.30] of the U.S. Geological Survey. It was felt that a chemical analysis of the highest quality by a great analytical chemist was needed for a calculation of this nature. Table 4.2 shows the chemical analysis of the Johanngeorgenstadt pitchblende.

The values of ϵ, p, f, η, and k_∞ thus calculated as a function of geological time are shown in Table 4.3. The result of the calculation was quite disappointing: it showed that the pitchblende was nuclear physically "stable" during the past 2800 million years. Moving back in time, the values of f and η steadily increased as the [235]U abundance increased, but the value of p decreased so that k_∞ also decreased after reaching a maximum value at about 1400 million years ago.

It occurred to the writer, however, that the above calculation was based upon an overly simplified model, in which it was assumed that an ore deposit was somehow instantaneously created in nature essentially in its present chemical state. Perhaps the uranium ores

4.30 W.F. Hillebrand, Bull. U.S. Geol. Survey 78:43 (1891); 90:23 (1892); 220:111−114 (1903). See also F.W. Clarke, The Data of Geochemistry, Washington Government Printing Office, 1924, p. 725

Table 4.3. The value of k_∞ for a Johanngeorgenstadt pitchblende as a function of geological time (P.K. Kuroda, J. Chem. Phys. 25:781 (1956))

Geological time (10^6 years ago)	0 (Present)	700	1000	1400	2100	2800
^{235}U enrichment (percent)	0.7	1.3	1.6	2.3	4.0	7.0
p	0.47	0.45	0.43	0.42	0.38	0.34
f	0.93	0.95	0.96	0.97	0.98	0.99
η	1.32	1.57	1.66	1.77	1.91	1.98
k_∞	0.58	0.67	0.69	0.72	0.71	0.67

should be treated as assemblages of enriched uranium (fuel), water (moderator) and neutron-absorbing impurities, in which the ratio (n) of water to uranium should be regarded as a variable.

The writer therefore considered that the formation of uranium ores represented the following sequence of events: an aqueous solution of uranium (^{235}U enriched) is gradually converted to an assemblage of uranium plus n moles of water and finally to an almost water-free uranium mineral. The values of p, f, η, and k_∞ were thus calculated as a function of n, assuming that Johanngeorgenstadt pitchblende formed 2100 million years ago. The results, shown in Table 4.4, clearly indicated that the system could have easily become nuclear physically "unstable", if the size of the assemblage was greater than, say, a thickness of a few feet.

The writer also noted in the 1956 paper that "the effect of the ground water or the vapor from the molten magma could have resulted in the formation of a nuclear physically 'unstable' assemblage of uranium plus n moles of water and such a mechanism might explain the fact that the ages of the large uranium deposits never exceed 2000 million years, or marked discrepancies exist between the ^{206}Pb/^{208}Pb age and the ^{207}Pb/^{208}Pb age of some of the uranium minerals."

In view of the results from these calculations, which seemed to strengthen the view that self-sustaining nuclear chain reactions could have occurred in nature during the geological history of the earth, it was felt to be worthwhile to look for the experimental evi-

Table 4.4. The water-uranium ratio n and the value of p, f, η, and k_∞ for a Johanngeorgenstadt pitchblende, 2100 million years ago (P.K. Kuroda, J. Chem. Phys. 25:781 (1956))

n	1/4	1/2	1	2	3	4	5	10
p	0.29	0.47	0.62	0.74	0.79	0.82	0.84	0.86
f	0.99	0.98	0.97	0.95	0.93	0.91	0.89	0.81
η	1.91	1.91	1.91	1.91	1.91	1.91	1.91	1.91
k_∞	0.55	0.88	1.15	1.34	1.40	1.42	1.43	1.33

Table 4.5. Calculated values of p, f, and k_∞ of the uraninites (P.K. Kuroda, J. Chem. Phys. 25:1295 (1956))

No.	Locality	p	f	k_∞	$\dfrac{Pb}{(U+0.36\,Th)}$
(1)	Placer de Guadalupe, Mexico	0.08	0.03	0.003	0.0046
(2)	Black Hawk, Colorado	0.30	0.90	0.36	0.009
(3)	Kirk Mine, Colorado	0.15	0.19	0.038	0.0115
(4)	Iizaka, Japan (clevite)	0.10	0.006	0.001	0.0136
(5)	Hale's Quarry, Connecticut	0.12	0.30	0.048	0.040
(6)	Blanchville, Connecticut	0.13	0.91	0.16	0.052
(7)	Boqueirao, Brazil	0.20	0.21	0.06	0.067
(8)	Johanngeorgenstadt, Saxony	0.47	0.93	0.58	0.084
(9)	Shinkolobwe, Katanga	0.08	0.20	0.021	0.084
(10)	Morogoro, East Africa	(0.22)[a]	0.08	(0.023)	0.088
(11)	Xique-Xique, Brazil	0.40	0.21	0.11	0.102
(12)	Gustav's Mine, Norway (brögerrite)	0.16	0.09	0.018	0.123
(13)	Lac Pied des Monts, Quebec	(0.22)[a]	0.14	(0.040)	0.148
(14)	Wilberforce, Ontario	0.16	0.02	0.004	0.157
(15)	Baringer Hill, Texas (nivenite)	0.33	0.007	0.003	0.163
(16)	Arendal, Norway (clevite)	(0.22)[a]	0.007	(0.002)	0.182
(17)	Great Bear Lake, Canada	0.37	0.08	0.041	0.202
(18)	Ingersoll Mine, South Dakota	0.09	0.08	0.010	0.226
(19)	Winnipeg River, Manitoba	(0.22)[a]	0.07	(0.020)	0.261
(20)	Sinyaya pala, Karelia, U.S.S.R.	0.29	0.024	0.009	0.300

[a] The values of p cannot be calculated, since the water contents of the minerals are unknown. An assumed value of $p = 0.22$ (an average value of p of the 16 samples of minerals) has been used for the calculation of k_∞

dence of the past existence of natural nuclear reactors. A search was made therefore for uranium minerals with large values of k_∞ and very old uranium-lead age, in which the values of k_∞ for a total of 20 samples of uraninite, pitchblende, bröggerite, nivenite, and clevite were calculated. Table 4.5 shows the results together with the Pb/(U + 0.36 Th) ratios of these minerals [4.31]. The approximate ages of the minerals listed in Table 4.5 can be obtained by multiplying the Pb/(U + 0.36 Th) ratios by 7.37×10^9 years.

The results from these calculations were disappointing: there appeared to be a trend that the greater the age of the mineral, the smaller the value of k_∞, although the minerals with small values of k_∞ are not necessarily old. It appeared as if minerals which are likely to contain detectable amounts of stable fission products other than fission krypton and fission xenon were absent in nature.

At about the same time, in 1956, research workers of the French Atomic Energy Commission were finding a new uranium ore deposit at Oklo, now in the Republic of Gabon, Africa. It so happened that the uranium ore from Oklo bore experimental evidences needed for the verification of the theory of natural reactors. If representative uranium specimens from Oklo were carefully examined mass-spectrometrically at that time, an anomaly in the isotopic composition of uranium indicating a depletion of ^{235}U in the ores should have been noted. However, it was 16 years later, in 1972, that such an anomaly was actually detected by French investigators.

4.31   P.K. Kuroda, J. Chem. Phys. 25:1295 (1956)

4.10. The Uranium-238 to -235 Ratio in Nature

It appeared to the writer that almost all of the scientists in the 1950's were of the opinion that the $^{238}U/^{235}U$ ratio is constant in nature. However, the published experimental data for the uranium isotopic ratio were far from being extensive. In 1956, Lounsbury [4.32] summarized the data available up to that time, including his new result on uranium extracted from Great Bear Lake Pitchblende, as shown in Table 4.6. In 1957, Senftle et al. [4.33] reported the $^{238}U/^{235}U$ ratio in thirteen uranium minerals, which were determined by the personnel of the Mass Assay Laboratory of Union Carbide Nuclear Company, Y-12 Plant at Oak Ridge, Tennessee. The results are shown in Table 4.7.

In 1960, Hamer and Robins [4.41] of the U.K. Atomic Energy Authority reported the results of a comparison of the isotopic compositions of uranium in twelve ore concen-

Table 4.6. The $^{238}U/^{235}U$ ratio in nature

Authors	Year	$^{238}U/^{235}U$
Nier [4.34]	1939	138.9 ± 1.4
Inghram [4.35]	1946	137.8
Fox et al. [4.36]	1946	137.0 ± 0.7
Chamberlain et al. [4.37]	1946	139
Bainbridge et al. [4.38]	1950	139.0 ± 1.4
Fleming et al. [4.39]	1952	138.0 ± 1.4
Hollander et al. [4.40]	1953	139.0 ± 1.4
Lounsbury [4.32]	1956	137.80 ± 0.14

Table 4.7. The $^{238}U/^{235}U$ ratio in various uranium minerals determined at the Mass Assay Laboratory of Union Carbide Company, as reported by Senftle et al. (1957)

	Minerals and ores	$^{238}U/^{235}U$
(1)	Sample G from Mineral Joe mine, Colorado	137.1 ± 0.35
(2)	Sample S from Mineral Joe mine, Colorado	137.7 ± 0.64
(3)	Sample J from Mineral Joe mine, Colorado	137.8 ± 0.31
(4)	Sample N from Mineral Joe mine, Colorado	137.7 ± 0.31
(5)	Uraninite, Happy Jack mine, Utah	137.8 ± 0.26
(6)	Oxidized ore, Happy Jack mine, Utah	137.8 ± 0.31
(7)	Carbonaceous ore, Temple Mountain, Utah	137.8 ± 0.31
(8)	Uraninite, Mi Vida mine, Utah	137.7 ± 0.31
(9)	Coffinite, Woodrow Pipe mine, New Mexico	137.6 ± 0.24
(10)	Coffinite, Poison Canyon mine, New Mexico	137.8 ± 0.31
(11)	Oxidized ore, black, J.J. mine, Paradox Valley, Colorado	137.8 ± 0.31
(12)	Uraninite, Joachimsthal	137.8 ± 0.28
(13)	Uraninite, Great Bear Lake, Canada	137.8 ± 0.25

4.32 M. Lounsbury, Can. J. Chem. 34:359 (1956)
4.33 F.E. Senftle, L. Stieff, F. Cuttitta and P.K. Kuroda, Geochim. Cosmochim. Acta 11:189 (1957)

Table 4.8. The constancy of the $^{238}U/^{235}U$ ratio in uranium ores according to Hamer and Robbins (1960)

	Ore deposit	Location	Maximum possible percentage deviation (D)
(1)	Wheal Edward	Cornwall, England	0.025
(2)	Magnesia Uranium Concentrate	Portugal	0.041
(3)	Shinkolobwe Pitchblende	Congo	0.046
(4)	Shinkolobwe "Ionex"	Congo	0.046
(5)	Mondola, Lot 1	Northern Rhodesia	0.028
(6)	Mindola, Lot 4 (high-copper content)	Northern Rhodesia	0.024
(7)	South African Concentrate	The Rand, South Africa	0.033
(8)	Rum Jungle	Northern Australia	0.038
(9)	Mary Kathleen	Queensland, Australia	0.043
(10)	Radium Hill	South Australia	0.033
(11)	Blind River	Ontario, Canada	0.028
(12)	Beaver Lodge	Alberta, Canada	0.028

trates by high-precision gas-source mass spectrometry. They expressed the data in the form of a maximum possible percentage deviation (D), as shown in Table 4.8.

In 1961, L.A. Smith [4.42] of the Oak Ridge Gaseous Diffusion Plant also reported the results of a comparison of the uranium isotopic ratios in fifteen ores. He expressed the results in the form of the ratio of the $^{235}U/(1 - ^{235}U)$ ratio of the sample to the $^{235}U/(1 - ^{235}U)$ ratio of the standard, as shown in Table 4.9.

In 1969 Szabo [4.43] summarized the $^{238}U/^{235}U$ ratios in various rocks and minerals, which have been measured by a number of investigators during the 1960's as shown in Table 4.10. In 1971, Rosholt and Tatsumoto [4.51] measured the isotopic composition of uranium in Apollo 12 lunar samples. Their results are shown in Table 4.11.

4.34 A.O. Nier, Phys. Rev. 55:150 (1939)
4.35 M.G. Inghram, Natl. Nuclear. Energy Ser. Div. II, 14:Chapt. V, 35 (1946)
4.36 M. Fox and B. Rustad, U.S. Atomic Energy Commission Rep. A-3828, 1946
4.37 O. Chamberlain, D. Williams and P. Yuster, Phys. Rev. 70:580 (1946)
4.38 K.T. Bainbridge and A.O. Nier, RNC-Nuclear Science Ser. Prelim. Rept. No. 9 (1950)
4.39 E.H. Fleming, A. Ghiorso and B.B. Cunningham, Phys. Rev. 88:642 (1952)
4.40 J.N. Hollander, I. Perlman and G.T. Seaborg, Rev. Mod. Phys. 25:469 (1953)
4.41 A.N. Hamer and E.J. Robbins, Geochim. Cosmochim. Acta. 19:143 (1960)
4.42 L.A. Smith, U.S. Atomic Energy Commission Report Number K-1462, January 19, 1961
4.43 B.J. Szabo, Etudes sur le Quaternaire dans le Monde, Union Internationale pur l'Etude du Quaternaire, VIIIe Congrès Inqua, Paris, 1969, p. 941
4.44 J.N. Rosholt, E.L. Garner and W.R. Shields, U.S. Geol. Survey Prof. Pap. 501-B:B87 (1964)
4.45 J.N. Rosholt, E.N. Harshman, W.R. Shields, and E.L. Garner, Econ. Geol. 59:570 (1964)
4.46 J.N. Rosholt and C.P. Ferreira, U.S. Geol. Survey Prof. Pap. 525-C:C58 (1965)
4.47 J.R. Dooley, H.C. Granger and J.N. Rosholt, Econ. Geol. 61:1362 (1966)
4.48 J.N. Rosholt, B.R. Doe and M. Tatsumoto, Geol. Soc. Am. Bull. 77:987 (1966)
4.49 B.L.K. Somayajulu, M. Tatsumoto, J.N. Rosholt and R.J. Knight, Earth Planet. Sci. Lett. 1:387 (1966)
4.50 B.J. Szabo and J.N. Rosholt, J. Geophys. Res. 74:3253 (1969)
4.51 J.N. Rosholt and M. Tatsumoto, Proceedings of the Second Lunar Science Conference, Vol. 2, The M.I.T. Press, 1971, pp. 1577–1584

Table 4.9. The constancy of the $^{235}U/^{238}U$ ratio as reported by L.A. Smith (1961)

Sample No.	Mineral	Mine Location	Ratio	95 percent Confidence Limits
542	Autunite in granite	Marysvale, Utah	1.002	± 0.0004
640–	Arsenical in sandstone	Camp Bird mine,	0.9999	± 0.0002
651		Temple Mountain, Utah	(0.9999	± 0.0002)
1235	Uraninite-fluorite-tyuyamunite in Todilto limestone	Haystack Mountain, McKinley Co., New Mexico	0.9993 (0.9994	± 0.0002 ± 0.0003)
1487	Uraninite in sandstone	Big Indian Wash, Utah	1.0001 (1.0000	± 0.0001 ± 0.0002)
1740	Low-vanadium oxide ore from sandstone, Petrified Forest, Chinle formation	Ramco No. 17 mine, Cameron, Arizona	0.9997	± 0.0002
1757	Uraninite-copper-vanadium in sandstone, Shinarump, Chinle formation	C-3 mine, Monument Valley, Utah, Arizona	0.9994 (0.9995	± 0.0002 ± 0.0002)
1762	Partly oxidized coffinite in sandstone, Brushy Basin, Morrison formation	Jackpile mine, Valencia County, New Mexico	0.9993 (0.9993	± 0.0002 ± 0.0002)
2294	Arsenical uraninite-coffinite in volcanic rocks	White King mine, Lakeview, Oregon	1.0002	± 0.0002
3033	Uraninite-copper in sandstone	White Canyon, Utah	0.9995 (0.9992	± 0.0002 ± 0.0002)
3034	Uraninite in sandstone, Upper Wind River formation	Aljob Claims, Natrona County, Wyoming	0.9996	± 0.0003
3035	Uranium-vanadium in Salt Wash sandstone, Morrison formation	Paradox D mine, Montrose Co., Colorado	0.9998	± 0.0003
3036	Uraninite (some coffinite) in metamorphic rocks	Schwartzwalder and Mena mines, Denver, Colorado	1.0001 (0.9999	± 0.0002 ± 0.0002)
3039	Schroeckingerite in mudstone, Lost Creek deposit	Sweetwater County, Wyoming	0.9995 (0.9995	± 0.0002 ± 0.0002)
Lot 19	Orange oxide	Belgian Congo	1.0000	± 0.0001
Lot 305	Orange oxide	Port Hope, Canada	0.9998	± 0.0001

Table 4.10. The $^{235}U/^{238}U$ ratio in natural materials, as summarized by Szabo (1969)

	Material	Variation of $^{235}U/^{238}U$ atomic ratio (percent)	Reference
(1)	Uranium ore deposits	< 0.05	Hamer and Robbins (1960) [4.41]
(2)	Granites and sandstones	< 0.5	Rosholt et al. (1964) [4.44]
(3)	Sandstones	< 0.5	Rosholt et al. (1964) [4.45]
(4)	Sandstone-type uranium deposits	< 0.5	Rosholt et al. (1965) [4.46]
(5)	Sandstone-type uranium deposits	< 0.5	Dooley et al. (1966) [4.47]
(6)	Soils	< 0.5	Rosholt et al. (1966) [4.48]
(7)	Basalts	< 0.5	Somayajulu et al. (1966) [4.49]
(8)	Marine shells	< 0.5	Szabo and Rosholt (1969) [4.50]

46

Table 4.11. Isotopic composition of uranium in Apollo 12 lunar samples, according to Rosholt and Tatsumoto (1971)

Sample	Rock type	U (ppm)	$^{234}U/^{238}U$ (activity ratio)	$^{238}U/^{235}U$ (atom ratio)
12013,10,09	Breccia	5.67	1.01	137.7
12013,10,42	Breccia	10.80	0.99	138.0
12013,10,45	Breccia	5.75	1.02	137.6
12034,16	Breccia	3.58	1.00	137.7
12033,53	Fines	2.67	0.99	137.7
12070,56	Fines (contingency)	1.64	1.01	137.8
12009.22	Crystalline rock	0.243	1.00	138.0
12021,122	Crystalline rock	0.261	1.00	137.8
12022,37	Crystalline rock	0.198	0.99	137.7
12035,10	Crystalline rock	0.199	0.99	137.6
12038,42	Crystalline rock	0.157	1.01	137.6
12052,66	Crystalline rock	0.365	1.00	137.8
12063,49	Crystalline rock	0.191	0.99	137.7
12064,21	Crystalline rock	0.278	1.01	137.9
Error range (%)				0.25

4.11. The Uranium-234 to -238 Ratio in Nature

In 1955, the Soviet investigators V.V. Cherdyntsev and co-workers [4.52] made the important discovery that the $^{234}U/^{238}U$ ratio does not remain constant in an extract from a mineral. For example, uranium in a zircon extract showed the $^{234}U/^{238}U = 1.8$ (curie/curie) and for uranium of surface waters, the observed ratio of $^{234}U/^{238}U$ equalled 7 to 8 (curie/curie).

According to Cherdyntsev [4.53], the process of radioactive decay of the heavy elements, such as uranium, produces a special case of variation of the isotopic ratio, which is not encountered in the case of stable isotopes. The decayed atoms, after recoil, pass into different sites of the crystalline lattice, where they reside in the form of foreign, generally weakly bound, inclusions. During the interaction of natural waters with minerals, the radioactive decay products pass into solution much more easily than do the isotopes, which did not undergo similar decay.

4.52 V.V. Cherdyntsev, P.I. Chalov, M.E. Khitrik, D.M. Mambetov and G.Z. Khaidarov, Proc. of the III Session of the Commission on Determination of the Absolute Age of Geological Formations, Acad. Sci. U.S.S.R., 1955

4.53 V.V. Cherdyntsev, Abundance of Chemical Elements, Translated by Walter Nichiporuk, The University of Chicago Press, 1961, p. 107

4.12. Discovery of the Oklo Reactor

On June 7, 1972, an anomaly in the uranium isotopic ratio of a sample of natural uranium was observed by H. Bouzigne, R.J.M. Boyer, C. Seyve, and P. Teulieres [4.54] at the French Atomic Energy Establishment at Pierrelatte. The ^{235}U abundance in this uranium was 0.7171 atom-percent, as against 0.7202 ± 0.0010 for normal natural uranium. During the months of June to August 1972, the following new facts emerged: (a) the anomalous ore was traced to Oklo in the Republic of Gabon, Africa, and it was found that, between December 1970 and May 1972, the ore which originated from Oklo was deficient in ^{235}U by a total amounting 200 kg of ^{235}U; (b) uranium with an unbelievably low isotopic abundance of 0.440 percent ^{235}U was discovered; and (c) fission-produced neodymium and samarium isotopes were found in the ore.

These developments led to the special announcement by the French Atomic Energy Commission on September 25, 1972: Uranium had been found, in the deposit at Oklo with an abnormal isotopic composition that led one to arrive at the conclusion that self-sustaining nuclear chain reactions had occurred on the earth approximately two billion years ago.

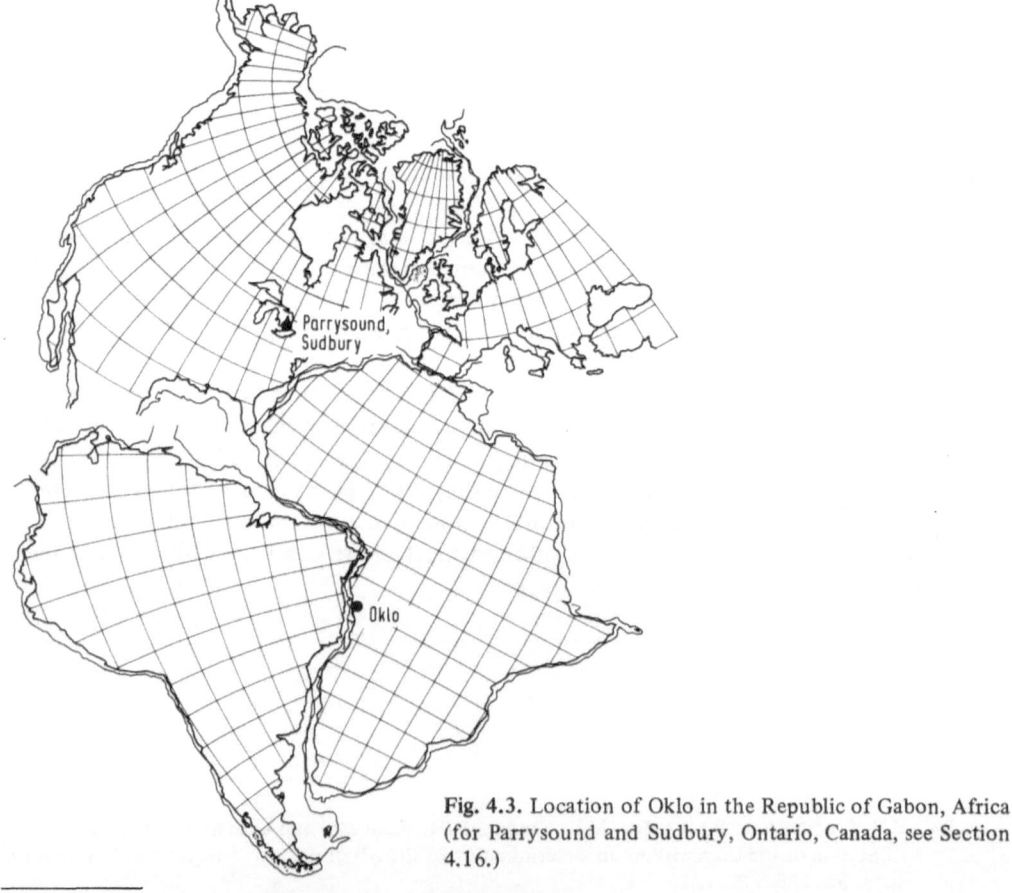

Fig. 4.3. Location of Oklo in the Republic of Gabon, Africa (for Parrysound and Sudbury, Ontario, Canada, see Section 4.16.)

4.54   H. Bouzigues, R.J.M. Boyer, C. Seyve and P. Teulieres, The Oklo Phenomenon, IAEA, Vienna, 1975, p. 237

The reaction site consited of several bodies of very rich uranium ore, and more than 500 tons of uranium had been involved in the reactions with a quantity of energy released equal to about 100×10^9 kW.h. The integrated flux at certain points exceeded 1.5×10^{21} n/cm^2, and samples have been found in which the concentration of the isotope ^{235}U was as low as 0.29 percent, as compared with 0.72 percent in natural uranium. Fig. 4.3 shows how the continents of Africa and America were joined together two billion years ago when the Oklo reactor was operating.

Results from the investigations of the Oklo reactor by French scientists have generally confirmed that the theory of natural reactors as it was originally envisioned by the writer in 1956 was essentially correct. But there were a couple of surprises. One was that the reactor seemed to have operated for a period as long as 600,000 to 1,500,000 years without destroying itself. Another surprise was that the formation of natural reactors appears to be closely related to the appearance of life on our planet earth.

According to Maurette [4.55], the high concentration of uranium found at Oklo resulted from a long series of repetitive fractionation processes, in which oxygen played a key role as an oxidizing agent. It is generally assumed that oxygen was injected into the earth's atmosphere only 2 billion years ago by a new generation of living organisms carrying out photosynthesis. Thus high-grade uranium ore deposit needed to trigger a nuclear chain reaction was probably never formed before 2 billion years ago and the occurrence of fossil reactors was probably limited to a relatively brief period of time ranging from 1 to 2 billion years ago.

The experimental data were reported in the fall of 1972 by three groups of researchers: Bodu and co-workers [4.56], Neuilly and co-workers [4.57] and Baudin and co-workers [4.58]. Table 4.12 shows the results reported by Neuilly and co-workers. The isotopic compositions of uranium and neodymium in the samples of Oklo-M and Oklo-310 are compared here with those in ordinary rocks. The isotopic compositions of neodymium found in these samples resemble those of fission-produced neodymium. Since ^{142}Nd is not formed by fission, it is possible to correct for the part due to the natural element and the isotopic compositions of neodymium thus corrected are compared with the known yields from the thermal neutron-induced fission of ^{235}U. As shown in Table 4.12, there is a good agreement between these values.

4.13. Promethium-147 in the Oklo Reactor

As shown in Table 4.12, the ^{149}Sm/^{147}Sm ratio in Oklo-M is abnormally low. The reason for this is that the cross-section for the neutron-capture reaction

$$^{149}\text{Sm}(n, \gamma)^{150}\text{Sm (stable)}$$

4.55 Michel Maurette, Annual Review of Nuclear Science, Vol. 26, 1976, p. 319
4.56 R. Bodu, H. Bouzigues, N. Morin and J.P. Pfiffelmann, C.R. Acad. Sc. Paris 275 D:1731 (1972)
4.57 M. Neuilly, J. Bussac, C. Fréjacques, G. Nief, G. Vendryes and J. Yvon, C. R. Acad. Sc. Paris 275 D:1847 (1972)
4.58 G. Baudin, C. Blain, R. Hagemann, M. Kremer, M. Lucas, L. Merlivat, R. Molina, G. Nief, F. Prost-Marechal, F. Regnaud and E. Roth, C. R. Acad. Sc. Paris 275 D:2291 (1972)

Table 4.12. Analyses of ore samples from the Oklo mines in Gabon, Africa. (Neuilly et al., C.R. Acad. Sci. Paris 275, Sér. D:1847 (1972))

Atomic concentration in the element (%)	Oklo M	Oklo 310	Natural element	Oklo M (*)	Oklo 310 (*)	^{235}U fission
^{235}U	0.4400 ±0.0005	0.592 ±0.001	0.7202			
^{142}Nd	1.38	5.49	27.11	0	0	0
^{143}Nd	22.1	23.0	12.17	22.6	25.7	28.8
^{144}Nd	32.0	28.2	23.85	32.4	29.3	26.5
^{145}Nd	17.5	16.3	8.30	18.05	18.4	18.9
^{146}Nd	15.6	15.4	17.22	15.55	14.9	14.4
^{148}Nd	8.01	7.70	5.73	8.13	8.20	8.26
^{150}Nd	3.40	3.90	5.62	3.28	3.46	3.12
Relative precision	2 to 3%	1%	0.2%			0.5%
^{151}Eu/^{153}Eu	0.145	0.852	0.916			2.58
^{140}Ce/^{142}Ce	1.57		7.99			1.06
^{149}Sm/^{147}Sm	~0.003		0.924			0.475

(*) Corrected for the portion due to the natural element

is 4,500 barns and most of the ^{149}Sm atoms were converted to the stable ^{150}Sm. The cross-section for the reaction

$$^{147}\text{Sm (n, }\gamma\text{) }^{148}\text{Sm (stable)}$$

is 60 barns and hence some of the ^{147}Sm atoms must have been converted to the stable ^{148}Sm, but the abundance of ^{147}Sm is obviously much less severely altered than that of ^{149}Sm. It is also interesting to note that ^{147}Sm is the daughter of ^{147}Pm:

$$^{147}\text{Pm} \xrightarrow[2.62\,\text{y}]{\beta^-} {}^{147}\text{Sm} \xrightarrow[1.07 \times 10^{11}\,\text{y}]{\alpha}$$

The presence of ^{147}Sm in the Oklo reactor can therefore be regarded as proof that element 61 once existed in nature in far greater quantities than the amount found in the uranium ore deposits today.

4.14. Plutonium-239 in the Oklo Reactor

As it was pointed out in Section 4.2., extremely small quantities of ^{239}Pu occur in uranium minerals. It is formed by the reaction:

$$^{238}\text{U} + \text{neutron} \rightarrow {}^{239}\text{U} \xrightarrow[\text{23.5-min}]{\beta} {}^{239}\text{Np} \xrightarrow[\text{2.35-day}]{\beta} {}^{239}\text{Pu} \qquad (4.4)$$

A far greater quantity of ^{239}Pu must have been produced in the Oklo reactor, since the neutron-flux there was much greater than in the ordinary environments. The ^{239}Pu thus produced must have decayed by emitting alpha particles:

$$^{239}\text{Pu} \xrightarrow[\text{24,000-year}]{\alpha} {}^{239}\text{U}$$

Now suppose that the duration of the natural reactor was comparable to or much longer than the half-life of ^{239}Pu. Much of the ^{239}Pu produced in the reactor must have undergone thermal neutron-induced fission and produced its own fission products:

$$^{239}\text{Pu} + \text{neutron} \rightarrow \text{fission products} + \text{neutrons},$$

while the reaction

$$^{235}\text{U} + \text{neutron} \rightarrow \text{fission products} + \text{neutrons}$$

was occurring in the reactor.

The fission products found in the Oklo reactor are therefore expected to be mixtures of the products of ^{235}U and ^{239}Pu fission. The fission yields from ^{235}U and ^{239}Pu are different from each other and are accurately known. It is therefore possible to determine the contribution from the ^{239}Pu fission by carrying out isotopic analysis of elements for which the plutonium gives fission yields quite different from those of uranium. Actually, the fast-neutron-induced fission of ^{238}U also contributes to the production of the fission products and it sometimes complicates the matter to some extent.

According to R. Hagemann and co-workers [4.59], the following relationships hold:

$$\frac{dN_{235}}{dt} = -N_{235}\, \sigma_{235}\, (1 - C)\, \phi$$

$$= -N_{235}\, \sigma_{235}\, \phi + N_{239}\, \lambda_{239} \qquad (4.5)$$

where C is a conversion factor, ϕ is the neutron flux and λ_{239} is the decay constant of ^{239}Pu, and

$$N_{239} = \frac{C N_{235}\, \sigma_{235}\, \tau}{\lambda_{239}\, \Delta t} \qquad (4.6)$$

where Δt is the duration of the reaction and τ is the fluence.

The percentage of the fissions due to ^{239}Pu can be written as

$$\beta = N_{239}\, \sigma_{f239} \left[\frac{1 - \alpha - \beta}{N_{235}\, \sigma_{f235}} \right] \qquad (4.7)$$

4.59 R. Hagemann, C. Devillers, M. Lucas, T. Lecomte, J.-C. Ruffenach, The Oklo Phenomenon, IAEA, Vienna, 1975, p. 415

where α and β are percentages of fissions due to ^{238}U and ^{239}Pu, respectively. It follows then

$$\Delta t = \frac{\sigma_{f239}\,\sigma_{235}}{\sigma_{f235}} \cdot \frac{1-\alpha-\beta}{\beta} \cdot \frac{C\tau}{\lambda_{239}} \tag{4.8}$$

The values of τ and C in equation (4.8) can be obtained relatively easily from the isotopic analysis of uranium and rare earths, especially neodymium and samarium. It is more difficult to obtain the values of α and β.

Hagemann and co-workers [4.59] determined these values from the observed ratio (R) of several fission products found in the ore samples. The ratios used for the calculations were:

^{150}Nd/(^{143}Nd + ^{144}Nd + ^{145}Nd + ^{146}Nd),
^{154}Sm/(^{147}Sm + ^{148}Sm),
(^{157}Gd + ^{158}Gd)/(^{155}Gd + ^{156}Gd),
^{110}Pd/^{105}Pd, ^{110}Pd/^{106}Pd, ^{104}Pd/^{101}Pd, and ^{104}Ru/^{102}Ru.

The following relationship holds between the fission product ratios and their fission yields (ρ' and ρ'') from ^{235}U, ^{238}U and ^{239}Pu:

$$R = \frac{\rho'_{235}\,(1-\alpha-\beta) + \rho'_{238}\,\alpha + \rho'_{239}\,\beta}{\rho''_{235}\,(1-\alpha-\beta) + \rho''_{238}\,\alpha + \rho''_{239}\,\beta} \tag{4.9}$$

The results obtained by Hagemann and co-workers [4.59] are shown in Table 4.13.

Table 4.13. Operating conditions of the Oklo Reactor (R. Hagemann and co-workers, The Oklo Phenomenon, IAEA, Vienna, 1975, p. 420)

Ore sample	α (percent)	β (percent)	Thermal neutron fluence, τ (n/cm^2)
KN50-3548	2.5 ± 1	4 ± 1	1.23 × 10^{21}
KN50-323	3 ± 1	3 ± 1	1.02 × 10^{21}
SC36-1413-3	3 ± 1	4 ± 1	1.32 × 10^{21}
SC36-1418	3.5 ± 1	4 ± 1	0.81 × 10^{21}

Conversion factor, C	Δt (Duration, years)
0.47	640,000
0.42	630,000
0.37	540,000
0.58	580,000

Table 4.14. Ruthenium isotopes in the uranium ores from Oklo (C. Fréjacques and co-workers, The Oklo Phenomenon, IAEA, Vienna, 1975, p. 517)

Sample	^{96}Ru	^{98}Ru	^{99}Ru	^{100}Ru	^{101}Ru	^{102}Ru	^{104}Ru
Natural	5.51	1.87	12.72	12.62	17.07	31.61	18.60
Fission (a)	— —	— —	34.66	— —	28.94	24.56	11.84
KN50-323	— —	— —	34.35	0.82	28.56	24.63	11.64
KN50-3548	— —	— —	31.35	0.95	29.48	25.87	12.35
SC36-1413-3	— —	— —	29.59	0.86	30.33	26.56	12.66
SC36-1418	— —	— —	27.70	0.82	30.97	27.33	13.18
SC36-1423-5	— —	— —	29.52	0.98	30.07	26.64	12.79

(a) 93% ^{235}U + 3% ^{238}U + 4% ^{239}Pu

4.15. Technetium-99 in the Oklo Reactor

Fréjacques and co-workers [4.60] measured the isotopic composition of ruthenium found in the Oklo deposits. As shown in Table 4.14, the relative abundances of ^{99}Ru, ^{101}Ru, ^{102}Ru and ^{104}Ru agree quite well with the calculated ratios for a mixture of fission products form 93 percent ^{238}U, 3 percent ^{235}U and 4 percent ^{239}Pu, but there is a small excess of ^{100}Ru. This is due to the fact that ^{99}Tc is converted to ^{100}Ru by neutron capture, while at the same time it decays to ^{99}Ru with a half-life of 2.13×10^5 years:

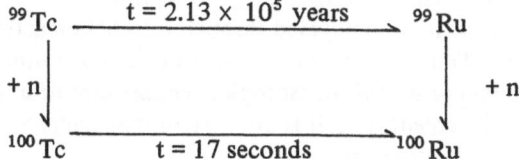

Hagemann and co-workers [4.59] have reported that the competition between the radioactive decay of ^{99}Tc to ^{99}Ru and neutron capture in ^{99}Tc and ^{99}Ru leading to ^{100}Ru enables one to determine the duration of the Oklo phenomenon. However, the relative migration of these two elements during the nuclear reaction sets a limit to the precision of the method.

The cross-section values are also not accurately known, but, as shown in Table 4.15, the duration of the nuclear reaction obtained by the ^{99}Tc method is considerably longer than the values obtained by the ^{239}Pu method. It is nevertheless important to note that the presence of ^{99}Ru and ^{100}Ru in the uranium ores from Oklo can be regarded as proof that element 43 once existed in nature in far greater quantities than the amounts found in the uranium ore deposits today.

4.59 R. Hagemann, C. Devillers, M. Lucas, T. Lecomte, J.-C. Ruffenach, The Oklo Phenomenon, IAEA, Vienna, 1975, p. 415
4.60 C. Fréjacques, C. Blain, C. Devillers, R. Hagemann and J.-C. Ruffenach, The Oklo Phenomenon, IAEA, Vienna, 1975, p. 509

Table 4.15. Influence of the neutron capture cross-sections for ruthenium-99 and technetium-99 on the calculation of the duration of the Oklo phenomenon (R. Hagemann and co-workers, The Oklo Phenomenon, IAEA, Vienna, 1975, p. 422)

| ^{99}Ru | | ^{99}Tc | | |
$\sigma_0{}'$ thermal (barns)	I resonance integral (barns)	$\sigma_0{}'$ thermal (barns)	I resonance integral (barns)	Duration (years)
5	195	19	340	1,400,000
5	195	21	360	1,700,000
5	195	17	320	1,150,000
5	245	19	340	3,500,000
5	145	19	340	900,000
5	170	18	330	1,000,000
5	145	17	320	740,000

4.16. Search for Additional Natural Reactors

Was Oklo a unique event in the history of the earth? According to Maurette [4.55], none of the characteristics of the Oklo deposit appear particularly unique and the Oklo phenomenon was probably not unique in the distant past. The only real unique feature of the Oklo deposit was the very careful routine analysis of uranium performed by H. Bouziques and co-workers [4.54]. The relative mass of uranium cycled through the chain reactions at Oklo has been evaluated at about 10^3. Thus the detection of fossil reactors in other uranium ores probably requires the frequent control of the isotopic compositions of uranium with an accuracy better than about 0.1 percent, and it is not certain that such good analyses have been and/or will be conducted in other uranium deposits.

According to G.A. Cowan [4.61], not all natural reactors would necessarily give rise to depleted uranium ores. If a natural reactor was able to form as late as 800 million years ago, when the relative abundance of ^{235}U was about 1 percent, it might have actually become a breeder reactor and the ^{235}U consumed in the reaction would have been more than replaced by new ^{235}U created by the decay of ^{239}Pu. Although the pitchblende deposit in the former Belgian Congo have now been mined out, precise isotopic analyses are available for a few samples of the ore and they appear to be slightly enriched in ^{235}U. Cowan also noted that uranium from the Colorado Plateau has a ^{235}U content slightly smaller than the world average and hence it is possible that a natural reactor once operated in this region.

As it has been mentioned in Section 4.3, Orr [4.7] once speculated that a critical uranium chain reaction might have taken place in thucholite from the Besner Mine, Parry Sound, Ontario, Canada (see Fig. 4.3 for the location of Parrysound). The isotopic compositions of xenon and krypton in several samples of thucholite from Parrysound have since been measured mass spectrometrically by Bogard et al. [4.62] and Kuroda and Sher-

4.61 G.A. Cowan, Scientific American 235:36 (1976)
4.62 D.D. Bogard, M.W. Rowe, O.K. Manuel and P.K. Kuroda, J. Geophys. Res. 70:703 (1965)

rill [4.63]. The relative abundances of the fissiogenic xenon and krypton isotopes released from the thucholite samples indicated, however, that they are primarily the product of ^{238}U spontaneous fission.

The isotopic compositions of xenon and krypton in several samples of granites have recently been measured mass spectrometrically by Kuroda et al. [4.64] and Hennecke et al. [4.65]. The results again indicated that they are primarily the product of ^{238}U spontaneous fission. A sample of Red Rock granite from Sudbury, Ontario, Canada (see Fig. 4.3 for the location of Sudbury), however, was found to contain an abnormally large amount of xenon. Moreover, a large fraction of the total xenon released from the Red Rock granite appeared to be a mixture of AVCC (average carbonaceous chondrite) xenon and the atmospheric xenon. According to Dietz [4.66], the Sudbury structure was produced by the impact of a 4 km-diameter asteroid in an event that released 3×10^{29} erg, or about a million megatons of energy, and formed a crater 30 miles in diameter. It is a remarkable coincidence that the event occurred at about the same time, when the natural reactors were in operation at Oklo. According to Bodu et al. [4.56], absolute chronology measurements of the Francevillian formations at Oklo indicated an age of 1,740 ± 20 m.y., while a whole rock Rb/Sr age determined on the micropegmatite of the Sudbury structure yielded a value of 1,720 m.y. [4.67].

4.63 P.K. Kuroda and R.D. Sherrill, Geochem. J. 11:9 (1977)
4.64 P.K. Kuroda, R.D. Sherrill and K.C. Jackson, Geochem. J. 11:75 (1977)
4.65 E.W. Hennecke, R.V. Ballad and O.K. Manuel, J. Inorg. Nucl. Chem. 40:1281 (1978)
4.66 R.S. Dietz, J. Geol. 72:412 (1964)
4.67 B.M. French, Science 156:1094 (1967); Bull. Volcanolog. 34-2:466 (1970); Geol. Assoc. Canada, Special Number 10:19 (1972)

5.
Synthesis of the Elements in Stars

*"The large amount of radioactive substances on the earth
and some stars may therefore be regarded as the results of
either somewhat sudden cooling of their mother rock or
the ejection from the interior of stars."*

Seitaro Suzuki, Proc. of the Physico-Math. Soc. Japan
3rd Ser. 13:277 (1931)

According to Burbidge, Burbidge, Fowler and Hoyle (B_2FH), most of the elements were
synthesized in the interior of stars by at least seven different types of nuclear processes.
Deuterium, lithium, beryllium and boron are very unstable, however, at the temperatures
of stellar interiors, so that they must have been produced in regions of low density and
temperature. The nature of the x-process, which is responsible for the synthesis of these
so-called "deficient" elements, has not yet been fully elucidated.

5.1. Discovery of Helium in the Sun

In 1868, an expedition was sent to India to observe an eclipse of the Sun. P.J.C. Janssen
first observed a yellow spectral line in the chromosphere of the Sun, and Frankland and
Lockyer concluded that it must be due to the presence in the solar atmosphere of an element unknown on the earth; they gave it the name "helium", to suggest its solar origin.
Helium remained as a hypothetical element which does not occur on the earth until 1895,
when it was discovered in uranium and thorium minerals by Sir William Ramsay.

In regard to the discovery of helium in radioactive minerals, Ramsay [5.1] stated in
1904: "It had been observed by Dr. Hillebrand, one of the chemists to the Geological
Survey of the United States, that certain minerals, notably those containing uranium,
gave off a gas when heated with dilute sulphuric acid. I was told recently that the investigators of this gas observed in its spectrum certain unknown lines; but they were deterred

---

5. 1   Sir William Ramsay, The Rare Gases of the Atmosphere, Nobel Lecture, December 12, 1904

by the criticism of their colleagues from attaching much importance to the observation. The spectrum contained the usual bands of nitrogen, and they contended themselves with chronicling this observation. The evolution of this gas from a rare mineral struck me as likely to afford a clue to the direction in which to seek for a compound of argon; for it appeared not unlikely that the gas examined by Hillebrand might turn out to be argon, mixed with nitrogen. But on examining the gas, a brilliant yellow line was observed, nearly, though not quite identical in position with the double line of sodium. Reference to a list of spectrum lines established the coincidence of the new line with a line which had been first observed in the solar spectrum by P.J.C. Janssen, during an eclipse of the sun, to observe which an expedition had been sent to India in 1868."

5.2. The Concept of Frozen Thermodynamic Equilibria

In 1922, Tolman [5.2] subjected the reaction

$$4 H = He \tag{5.1}$$

to a complete thermodynamic treatment and found that hydrogen should combine practically completely to form helium at all temperatures below a million degrees and pressures above 10^{-100} atmospheres.

Suzuki [5.3], in 1927, arrived at a different conclusion, however, by extending the theory of Saha [5.5] on the ionization of atoms on the surface of stars to the problem of thermal dissociation of the atomic nuclei in the core of stars. According to Suzuki's calculations, helium nuclei dissociated into protons only when the temperature was raised to about 2×10^9 degrees.

Later, Suzuki [5.4] proceeded to investigate the thermodynamic equilibria in terms of the classical Gibbs grand canonical ensemble [5.6] and also considered the dissociation of heavy elements, such as the pair of radium and helium nuclei in stars. He went on to attempt to explain the appearance of Novae based on his temperature-pressure-dissociation diagram and stated: "When the temperature is raised to that of the state where the resonance-disintegration manifests the conspicuous activity – about 10^9 degrees, the thermodynamic equilibrium is attained quickly. In this stage the thermodynamical equation shows also the maximum abundance of radium. The large amount of radioactive substances on the earth and some stars may therefore be regarded as the results of either somewhat sudden cooling of their mother rock or the ejection from the interior of stars."

Suzuki was apparently under the influence of the noted Japanese physicist Hantaro Nagaoka [5.7, 5.9], who in 1903 compared the atom to the planet Saturn, where stability is maintained by the attraction of the heavy central body for the lighter particles in the surrounding rings, and stated: "The present case (i.e., the atom) will evidently be approxi-

---

5. 2   R.C. Tolman, J. Am. Chem. Soc. 44:1902 (1922)

5. 3   S. Suzuki, Proc. Imperial Acad. Japan III, 10:650 (1927)

5. 4   S. Suzuki, Proc. of the Physico-Math. Soc. Japan: 3rd Series 10:166 (1928); 11:119 (1929); 13:277 (1931)

5. 5   M.N. Saha, Zeits. f. Physik 6:40 (1921)

5. 6   J. Willard Gibbs, Collected Works, Yale University Press, New Haven, 1948

5. 7   H. Nagaoka, Phil. Mag. 7:445 (1904)

mately realized if we replace these satellites by negative electrons and the attracting center by a positively charged particle." According to Suzuki [5.8], Nagaoka put forward the following view on the Universe in 1916 or 1917: "In the beginning there were groups of fine dust particles; due to the universal gravitational attraction, they became small stars, which in turn attracted each other and became big stars; they contracted gradually and became hotter and brighter; they exploded and flew away under the great light pressure; they gathered by the attraction and repeated the process of explosion, but finally all became interstellar dust particles, out of which small and large stars were formed again."

In 1931, Farkas and Harteck [5.10] reported that the equilibrium distribution of nuclei was established at a high temperature (about 10^9 K) and at densities of about 10^5 g/cm^3, and was then frozen in by the cooling of the stellar body. Pokrowski [5.11], also in 1931, pointed out, however, that there was a large discrepancy between the observed relative abundances of the heavy elements and those calculated by a simple equilibrium theory.

The basic principles of the equilibrium theory depend on the classical Gibbs grand canonical ensemble [5.6], the method of Darwin and Fowler [5.12] and the analog of Saha's ionization equation [5.5]. A most complete calculation of the relative abundances of the elements based on the equilibrium theory was performed by Klein, Bestow and Treffenberg [5.13] in 1947. The general agreement with increasing atomic weight up to about A = 40 was satisfactory, but the abundance peak near iron could not be reproduced by the theory. For the heavier elements, the computed abundances were extraordinarily small compared to the observational data. These results obtained during the 1940's were reviewed by Alpher and Herman [5.14] in 1950.

5.3. Deficient Elements

In 1926, V.M. Goldschmidt [5.15] first noted that the light elements lithium, beryllium and boron are "deficient" in nature and suggested that this deficiency must have a common cause for all three elements (mass numbers 6, 7, 9, 10 and 11) which might be sought in some instability in certain nuclear processes. In 1928, Gamow [5.16], and independently Gurney and Condon [5.17] studied the process of alpha-decay based on the wave mechanics in order to explain the experimentally found relationship between the decay constant and the energy of the alpha-particles: the so-called Geiger-Nuttall rule [5.18]. Atkinson and Houtermans [5.19] then calculated the probability for the penetration of the protons into the nucleus of light elements at stellar temperatures ($1 - 4 \times 10^7$ degrees). The results indicated that the deficient elements are rapidly destroyed in stars at these temperatures (see Chapter 7, Section 7.1, Origin of Lithium, Beryllium and Boron).

5. 8 S. Suzuki, Butsurigakushi-Kenkyu (Research in History of Physics) Vol. 3, No. 2, 1966 (in Japanese)
5. 9 Samuel Glasstone, Sourcebook on Atomic Energy, D. Van Nostrand, New York, 1950, p. 80
5.10 L. Farkas and P. Harteck, Naturwiss. 19:705 (1931)
5.11 G.I. Pokrowski, Physik. Z. 32:374 (1931)
5.12 R.H. Fowler, Statistical Mechanics, Cambridge University Press, London, 1929
5.13 O. Klein, G. Beskow and L. Treffenberg, Arkiv. f. Mat. Astr. O. Fys. 33B, No. 1, 1946; 34A, No. 13, No. 17 and No. 19 (1947)
5.14 R.A. Alpher and R.C. Herman, Rev. Mod. Phys. 22:153 (1950)

5.4. The Rate of Thermonuclear Reactions

In 1938, Gamow [5.20] reported that in order to initiate the formation of heavier elements inside a star it is necessary that some of these thermonuclear reactions between light elements should produce an appreciable number of neutrons, which would then be radiatively captured by heavier nuclei. This would result in a continuous buildup of heavy elements.

It seemed that the only way in which neutrons could be produced at comparatively low energies of collision ($4 \times 10^{7\circ}$ C corresponding to an energy of 10 keV) was by the preliminary formation of deuterons, which undergo in mutual collision the following transformations:

$$_1H^2 + {}_1H^2 \rightarrow {}_1H^3 + {}_1H^1 ; {}_1H^2 + {}_1H^2 \rightarrow He_3 + {}_0n^1 \tag{5.2}$$

There are about equal probabilities that the above two reactions will occur.

Gamow then considered the following two chains of thermonuclear reactions, which would produce deuterium:

$$\begin{aligned}
&_2He^4 + {}_1H^1 \rightarrow {}_3Li^5 + h\nu, \\
&_3Li^5 \rightarrow {}_2He^5 + \beta^+, \\
&_2He^5 + {}_1H^1 \rightarrow {}_2He^4 + {}_1H^2 .
\end{aligned} \tag{5.3}$$

and

$$\begin{aligned}
&_2He^4 + {}_2He^4 \rightarrow {}_4Be^8 + h\nu, \\
&_4Be^8 + {}_1H^1 \rightarrow {}_5B^9 + h\nu, \\
&_5B^9 \rightarrow {}_4Be^9 + \beta^+, \\
&_4Be^9 + {}_1H^1 \rightarrow 2{}_2He^4 + {}_1H^2
\end{aligned} \tag{5.4}$$

It is interesting to note here that in both chains the original $_2He^4$ nuclei which entered the reaction are always set free at the end; helium acts as a catalyst for the process of transmutation of ordinary hydrogen into heavy hydrogen.

Both possibilities (5.3) and (5.4) rest, however, on the stabilities of He^5 and Be^8. Although there was this uncertainty, Gamow considered that the deuterium nuclei produced in these chains would react according to equation (5.2). There will be a large number of neutrons thus produced in a star. These neutrons will be captured by heavier elements and will increase the mass of the heavy elements.

In 1938, G. Gamow and E. Teller [5.21] derived the equations which enable one to obtain the rate of selected thermonuclear reactions. They considered a gas consisting of two elements with atomic number Z_1 and Z_2 and atomic masses m_1 and m_2. If x and y are the relative amounts (by weight) of two components, and ρ and T the total density and temperature, the number of collisions between two nuclei of the two different kinds with the collision energy in the interval E and E + dE is given by

5.15   V.M. Goldschmidt, Gerlands Beitr. zur Geophys. 15:38 (1926)
5.16   G. Gamow, Z. f. Physik 51:204 (1928)
5.17   R.W. Gurney and E.U. Condon, Nature 122:439 (1928); Phys. Rev. 33:127 (1929)
5.18   H. Geiger and J.M. Nuttall, Phil. Mag. 22:613 (1911); 23:439 (1912)
5.19   R. d'E. Atkinson and F.G. Houtermans, Z. f. Physik 54:656 (1929)

$$dN = \frac{4xy\rho\sigma}{(2\pi)^{1/2}\ m_1\ m_2\ m^{1/2}\ (kT)^{3/2}}\ e^{-E/kT}\ E\ dE \tag{5.5}$$

where

$$m = \frac{m_1\ m_2}{m_1 + m_2} \tag{5.6}$$

is the reduced mass and σ the effective cross section.

Using Gamow's penetration formula, the effective cross section can be written as

$$\sigma \cong \frac{\Lambda^2}{4\pi}\ \exp\left[\frac{-2\,\pi\,e^2\,m^{1/2}\,Z_1 Z_2}{\hbar(2E)^{1/2}} + \frac{4e\,(2m\,Z_1 Z_2 r_0)^{1/2}}{\hbar}\right] \frac{\Gamma_\gamma\,m\,r_0^2}{\hbar} \tag{5.7}$$

where

$$\Lambda = \frac{2\,\pi\,\hbar}{(2\,m\,E)^{1/2}} \tag{5.8}$$

is the de Broglie wave-length and Γ_χ is the emission probability of γ-rays.

Substituting (5.7) into (5.5), Gamov and Teller obtained an expression possessing a sharp maximum at

$$E \sim (\pi\,e^2\,m^{1/2}\,Z_1 Z_2 kT)^{2/3}/(2^{1/2}\,\hbar)^{2/3} \tag{5.9}$$

the breadth of which was

$$\Delta E \sim (8/3\ kt)^{1/2}\ (2\,\pi\,e^2\,m^{1/2}\,Z_1 Z_2 kT^{1/3})/(2^{1/2}\,\hbar)^{1/3}$$

For the total number of captures per unit mass they obtained:

$$N \cong \frac{\pi^{5/6}}{3^{1/2}}\ \frac{e^{2/3}\,\hbar^{2/3}\,Z_1^{1/3}\,Z_2^{1/3}\,r_0^2\,\Gamma_\gamma}{m_1\ m_2\ m^{1/3}}$$

$$\exp\left[\frac{4e\,(2m\,Z_1 Z_2 r_0)^{1/2}}{\hbar} - 3\left(\frac{\pi^2\,e^4\,m\,Z_1^2 Z_2^2}{2\hbar^2\ kT}\right)^{1/3}\right]\frac{x\,y\,\rho}{(kT)^{2/3}} \tag{5.10}$$

If the above formula is applied to the d-d reaction:

$$_1H^2 + _1H^2 = {}_2He^3 + {}_0n^1$$

which is evidently the fastest of all thermonuclear reactions, we have

$$Z_1 = Z_2 = 1$$
$$m_1 = m_2 = 2$$

and

$$x = y = 1/2$$

Assuming that $\rho = 1/7$ g/cm^3, which represents the density of liquid deuterium, the results shown in Table 5.1 were obtained for the energy production.

Table 5.1. The energy production from the d-d reaction (Gamow and Critchfield, 1949) [5.22]

Temperature (°C)	Energy Production (cal./g. sec.)
5×10^5	10^{-7}
1×10^6	300
3×10^6	3×10^7
1×10^7	3×10^{12}

5.5. The C-N Cycle and the Proton-Proton Chain

Weizsäcker [5.23], in 1937, and Bethe and Critchfield [5.24], in 1938, studied the following reaction for the source of energy production in the sun and stars in general:

$$^1H + {}^1H = {}^2D + \epsilon^+ + \nu \tag{5.11}$$

where ϵ^+ = positron and ν = neutrino.

Although the proton-proton reaction (5.11) predicts the correct energy production in the sun, it has rather weak dependence on temperature. It was known, however, that the observed energy production in stars increased rapidly with increasing mass and hence Bethe felt that there must exist nuclear reactions which are more strongly dependent on temperature. These reactions must involve heavier nuclei.

Bethe [5.25, 5.26] examined the reactions between H and He, Li, Be and B. Reactions of H with D, Li, Be, and B are all very fast at the central temperature of the sun. This means that the partner of H is very quickly used up in the process. In fact, and just because of this reason, D, Li, Be and B are "deficient" on earth and in the stars, and they can not be important sources of energy.

Bethe found, however, that the next element, carbon, undergoes a cycle of reactions as follows:

5.20 G. Gamow, Phys. Rev. 53:595 (1938)
5.21 G. Gamow and E. Teller, Phys. Rev. 53:608 (1938)
5.22 G. Gamow and C.L. Critchfield, Theory of Atomic Nucleus and Nuclear Energy Source, Oxford University Press, 1949, p. 264
5.23 C.F. v. Weizsäcker, Phys. Z. 38:176 (1937)
5.24 H.A. Bethe and C.L. Critchfield, Phys. Rev. 54:248 (1938)
5.25 H.A. Bethe, Phys. Rev. 55:436 (1939)
5.26 H.A. Bethe, Energy Production in Stars, Nobel Lecture, December 11, 1967; Science 161:541 (1968)

$$
\begin{array}{lll}
^{12}\text{C} + {}^{1}\text{H} = {}^{13}\text{N} + \gamma & \text{(a)} & \\
\qquad\quad {}^{13}\text{N} = {}^{13}\text{C} + e^{+} + \nu & \text{(b)} & \\
^{13}\text{C} + {}^{1}\text{H} = {}^{14}\text{N} + \gamma & \text{(c)} & \text{(5.12)} \\
^{14}\text{N} + {}^{1}\text{H} = {}^{15}\text{O} + \gamma & \text{(d)} & \\
\qquad\quad {}^{15}\text{O} = {}^{15}\text{N} + e^{+} + \nu & \text{(e)} & \\
^{15}\text{N} + {}^{1}\text{H} = {}^{12}\text{C} + {}^{4}\text{He} & \text{(f)} &
\end{array}
$$

Reactions (a), (c), and (d) are radiative captures; the proton is captured by the nucleus and the energy is emitted in the form of gamma rays. The gamma ray energies are then quickly converted into thermal energy of the gas.

Reactions (b) and (e) are simply spontaneous beta decays. Their lifetimes are 10 minutes and 2 minutes, respectively, and hence negligible in comparison with the lifetimes of the sun or the stars.

Reaction (f) is very interesting, because it closes the cycle. The ^{12}C which we started from is reproduced; carbon is only used as a *catalyst*.

The summation is a combination of four protons and two electrons to form one ^{4}He nucleus: the electrons are used to annihilate the positrons emitted in reactions (b) and (c).

In the C-N cycle, two neutrinos are emitted and they carry away about 2 MeV energy together. The rest of the energy, about 25 MeV per cycle, is released usefully *to keep the sun warm*.

Since the C-N cycle involves nuclei of relatively high charge, it has a strong temperature dependence. The reaction (d) is the slowest of the cycle and therefore determines the rate of energy production. According to Bethe, it goes about as T^{24} near solar temperature, and this is sufficient to explain the high rate of energy production in massive stars. The carbon nitrogen cycle was also discovered independently by Weizsäcker [5.27], who recognized that this cycle consumes only the most abundant element, hydrogen.

The proton-proton reaction (5.11) also forms a cycle:

$$
\begin{array}{l}
^{1}\text{H} + {}^{1}\text{H} = {}^{2}\text{D} + e^{+} + \nu \\
^{2}\text{D} + {}^{1}\text{H} = {}^{3}\text{He} \\
^{3}\text{He} + {}^{3}\text{He} = {}^{4}\text{He} + 2{}^{1}\text{H} \\
\hline
4{}^{1}\text{H} = {}^{4}\text{He} + \text{energy}
\end{array}
\qquad (5.13)
$$

The net result is again a combination of four protons and two electrons to form one ^{4}He nucleus and about 26 MeV of energy is released per cycle.

5.6. Synthesis of the Elements in a Neutron-Rich Environment

During the early 1940's, Cherdyntsev [5.28] proposed a hypothesis of the formation of atomic nuclei under the conditions of thermodynamic equilibrium of a neutron environment. His assumption was used subsequently by Mayer and Teller [5.29] in the hypothe-

5.27 C.F. v. Weizsäcker, Phys. Z. 39:633 (1938)
5.28 V.V. Cherdyntsev, DANS SSR (Repts. Acad. Sci. U.S.S.R.) 25:19 (1941); Abundance of Chemical Elements, translated by W. Nichiporuk, The University of Chicago Press, 1961, p. 242
5.29 M.G. Mayer and E. Teller, Phys. Rev. 76:1226 (1949)

sis of the fission of a "polyneutron" nucleus and also the idea of the formation of atomic nuclei during the interaction of neutrons with matter was utilized, in the general form, in the theory of Alpher, Bethe and Gamow [5.30].

According to Cherdyntsev, the relative concentrations of the nucleus with atomic weight A under conditions of thermodynamic equilibrium of the system can be written in the form

$$Q = c_1 \pm c_2 A - c_3 A^{2/3} - c_4 Z^2 A^{-1/3} \qquad (5.14)$$

where c_1, c_2, c_3, and c_4 represent certain coefficients which include thermodynamic parameters.

The expression for the neutron-rich nucleus is

$$Q = c_1 \pm c_2 A - c_3 A^{2/3} \qquad (5.15)$$

The equation (5.15) differs from equation (5.14) only in that it does not contain the Coulomb energy term and the coefficients have different values. The presence of the Coulomb energy term in equation (5.14) causes a rapid fall in the abundances of the heavy elements.

5.7. The Big-Bang Theory of Gamow

In 1946, Gamow [5.31] pointed out that various species must have originated not as a result of an equilibrium corresponding to a certain temperature and density, but rather as a consequence of a continuous buildup process arrested by a rapid expansion and cooling of the primordial matter. He imagined the early stage of matter as a highly compressed neutron gas (*overheated neutral nuclear fluid* = "ylem") which started decaying into protons and electrons when the gas pressure fell down as the result of universal expansion. The radiative capture of the remaining neutrons by the newly formed protons leads first to the formation of deuterium nuclei, and the subsequent neutron captures result in the building up of heavier and heavier nuclei.

Due to the comparatively short time allowed for this buildup process, the synthesis of heavier nuclei must have proceeded just above the upper fringe of the stable elements, and the present frequency distribution of various atomic species was attained only somewhat later as the result of adjustment of their electric charges by beta-decay.

Alpher, Bethe and Gamow [5.30], in 1948, thus argued that the observed slope of the abundance curve must not be related to the temperature of the original neutron gas, but rather to the time period permitted by the expansion process.

It follows then that the individual abundances of various nuclear species must depend not so much on their intrinsic stabilities (*mass defect*) as on the values of their neutron capture cross sections.

5.30   R.A. Alpher, H. Bethe and G. Gamow, Phys. Rev. 73:803 (1948)
5.31   G. Gamow, Phys. Rev. 70:572 (1946)

The equation governing such a buildup process can be written in the form:

$$dn_i/dt = f(t) \, (\sigma_{i-1} \, n_{i-1} - \sigma_i \, n_i) \tag{5.16}$$

where n_i and σ_i are the relative numbers and capture cross sections for the nuclei of atomic weight i, and where $f(t)$ is a factor characterizing the decrease of the density with time.

According to Hughes [5.32], the neutron capture cross sections of various elements (for neutron energies of about 1 MeV) increase exponentially with atomic number halfway up the periodic system, remaining approximately constant for heavier elements.

By integrating equation (5.16) and using these cross sections, one finds that the relative abundances of various nuclear species decrease rapidly for the light elements and remain approximately constant for the elements heavier than silver.

In order to fit the calculated curve with the observed abundances [5.33] it is necessary to assume the integral of ρ_n dt during the buildup period to be equal to

$$5 \cdot 10^4 \text{ g sec/cm}^3.$$

On the other hand, according to the relativistic theory of the expanding universe [5.34], the density dependence on time is given by

$$\rho \cong 10^6/t^2 \tag{5.17}$$

Since the integral of this expression diverges at $t = 0$, it is necessary to assume that the buildup process began at a certain time t_0, satisfying the relation

$$\int_{t_0}^{\infty} (10^6/t^2) \, dt \cong 5 \cdot 10^4 \tag{5.18}$$

which gives us

$$t_0 = 20 \text{ sec} \tag{5.19}$$

and

$$\rho_0 = 2.5 \cdot 10^5 \text{ (g sec/cm}^3) \tag{5.20}$$

5.8. The Polyneutron Hypothesis of Mayer and Teller

In 1949, Mayer and Teller [5.29] pointed out that the abundances of elements and of isotopes indicated that heavy and light elements were produced by different processes, and discussed the origin of heavy elements in some detail. They assumed that the heavy ele-

5.32   D.J. Hughes, Phys. Rev. 70:106(A) (1946)
5.33   V.M. Goldschmidt, Geochemisches Verteilungsgesetz der Elemente und der Atom-Arten, IX, Oslo, Norway, 1938
5.34   R.C. Tolman, Relativity, Thermodynamics and Cosmology, Oxford, England, 1934

ments were formed by a "fission" process from a neutron-rich nuclear fluid. At the time of production of the heavy nuclei, the proportion of neutrons must have considerably exceeded that which is now present in nuclei for the following reasons: (a) without such a neutron-excess it is not possible to understand that the heavy isotopes of heavy elements are much more abundant than the lightest isotopes; and (b) in the absence of a neutron excess it is very hard to find any method by which the heavy nuclei could be built up at all.

It is interesting to note that Mayer and Teller assumed the initial presence of elements much heavier than uranium; up to $Z = 137$. They excluded the elements with Z greater than 137, however, since in a Coulomb field with Z greater than 137 the electron orbits with $j = 1/2$ can no longer exist outside of the nucleus. They also discussed a model of a nucleus which can serve as the starting point for the break-up process. An assembly of neutrons may form a *nuclear fluid* which will not spontaneously disintegrate into neutrons and the only limitation imposed on the size of this *polyneutron* may be that its total mass should not exceed the mass of a star.

5.9. The Proton-Neutron Ratio Prior to the Big-Bang

In 1950, C. Hayashi [5.35] reported that under the original assumption of Gamow that „ylem" consists solely of neutrons, it is difficult to explain the fact that the buildup processes of the elements jump over the crevasses of unstable mass numbers 5 and 8. He felt that the existence of an appreciable amount of original protons may relieve this situation.

According to Hayashi, at early stages of high temperatures ($kT > mc^2$, where m is the electron mass) in the expanding universe before the formation of the elements, induced beta-processes caused by energetic electrons, positrons, neutrinos and anti-neutrinos, in addition to the natural decay of neutrons, such as

$$
\begin{aligned}
n + e^+ &\rightleftharpoons p + a\nu \\
n + \nu &\rightleftharpoons p + e^- \\
n &\rightleftharpoons p + e^- + a\nu
\end{aligned}
\tag{5.21}
$$

must have occurred, and, since their rates are faster at higher temperatures, these reactions must have influenced the initial proton-neutron concentration ratio.

At still higher temperatures ($kT \geqslant \mu c^2$, where μ is the mass of meson), the neutron-proton conversion process induced by mesons is expected to be much more rapid.

Based on the relativistic theory of the expanding universe, Hayashi was able to calculate the neutron/proton ratio to be, for example, 1 : 4 at $2 \cdot 10^{10}$ °K. From this initial ratio, helium will be produced by the processes, such as:

$$
\begin{aligned}
n + p &\rightarrow {}^2H, \\
{}^2H + {}^2H &\rightarrow {}^3H + p, \\
{}^3H + {}^2H &\rightarrow {}^4He + n,
\end{aligned}
\tag{5.22}
$$

or

$$
\begin{aligned}
n + p &\rightarrow {}^2H \\
{}^2H + n &\rightarrow {}^3H \\
{}^3H + p &\rightarrow {}^4He
\end{aligned}
\tag{5.23}
$$

5.35   C. Hayashi, Progress of Theoret. Phys. 5:224 (1950)
5.36   F. Hoyle, Astrophys. J. Supplement I:121 (1954)

Consequently, the hydrogen-to-helium abundance ratio (atom/atom) resulting from the initial n/p ratio = 1/4 becomes:

H/He = 6 (5.24)

Hayashi stated that this H/He ratio is in agreement with observed values in stellar atmospheres and meteorites ranging from 5 to 10.

5.10. Theories on the Evolution of Stars

In 1954, F. Hoyle [5.36] reported that the abundances of the chemical elements over the portion of the periodic table from carbon to nickel are consistent with the view that the elements originate at the high temperatures that probably occur in the interiors of certain types of stars. The temperatures under consideration lie considerably above the temperatures occurring in ordinary main-sequence stars and are to be thought of as temperatures attained in a collapsing sequence following the exhaustion of hydrogen in luminous stars. Hoyle felt that perhaps the best observational indication of the existence of collapsing stars comes from the faint blue stars present in the globular clusters and also in regions of the sky near the galactic pole. These are the so-called type II stars.

According to Hoyle, supernovae of Type I occur among the Type II stars. He suggested that *it is by this process that elements other than hydrogen and helium are built up and distributed in the universe.* Fig. 5.1 shows the general cosmological framework assumed by Hoyle.

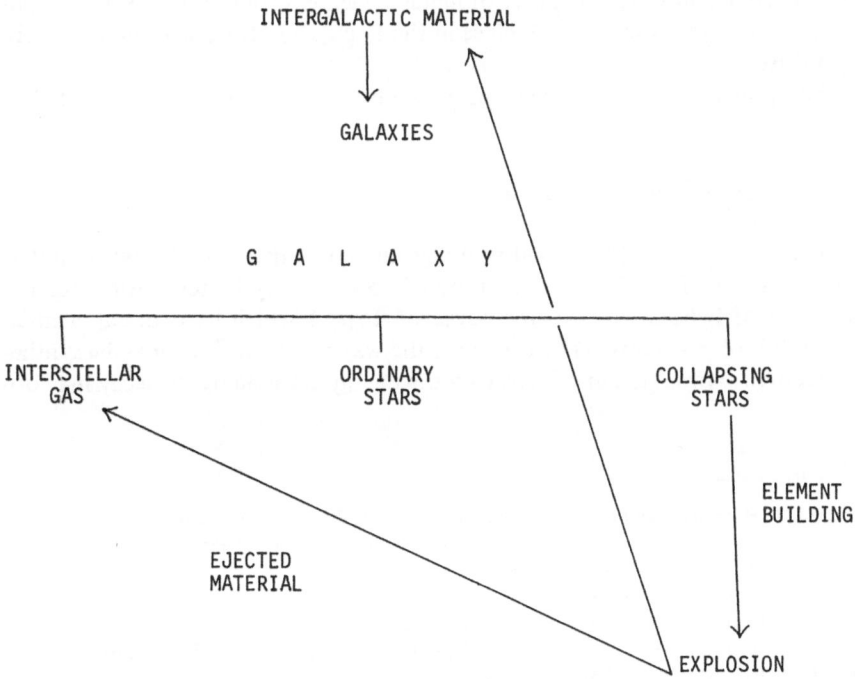

Fig. 5.1. The general cosmological framework assumed by Hoyle, Astrophysical J. Supplement I:121 (1954)

The galaxies, which are taken as forming from an extra-galactic cloud, are shown as consisting of interstellar gas, ordinary noncollapsing stars, and collapsing stars. Heavy elements are generated and scattered into space by the latter, or by a proportion of them. The scattering process may lead either to the heavy elements being trapped inside interstellar gas or to their escaping entirely from the parent-galaxy into the extragalactic medium. The velocities of ejection from many exploding stars — the novae and supernovae in particular — considerably exceed the velocities of escape from the galaxies.

The enrichment in this way of the heavy-element component of an extragalactic gas cloud means that even if the cloud were initially pure hydrogen, only the first galaxies formed could be of pure hydrogen. Subsequent galactic condensations must contain heavy elements ejected from the "first" galaxies. Strictly speaking, the concept of "first" galaxies can be applied only in cosmologies that assign a finite age to the universe. Thus, according to Hoyle, in the *steady-state* theory of the universe there are no "first" galaxies, since, in this theory, the universe has an infinite past. Consequently, at all times the extragalactic material must contain heavy elements that were ejected from previously existing galaxies. Accordingly, all galaxies should possess heavy elements at the time of their birth.

In 1955, Taketani, Hatanaka and Obi [5.37] proposed a scheme for the whole evolutionary process of stars. They reported that the two populations of stars (Type I and Type II) are made up of fundamentally different kinds of stars and these differences can not be accounted for by a simple evolutionary process of stars. Hoyle and Schwarzschild [5.38], in 1955, applied models with composite structures to explain the evolutionary track for Population II stars. Roy [5.39] and Tayler [5.40] investigated the evolution of Population I stars. The whole evolutionary process including the stars of Populations I and II and the interstellar gas was suggested by Schwarzschild and Spitzer [5.41], who attributed the difference in the abundances of metallic elements in the stars of two populations to the nuclear reaction which took place in the Population II stars in the very early stage of the universe.

Fig. 5.2 shows the evolutionary scheme proposed by Taketani and co-workers [5.37].

5.11. Supernovae and Californium 254

In 1956, Baade and co-workers [5.42] and Burbidge and co-workers [5.43] suggested that the spontaneous fission of ^{254}Cf with a half-life of 55 days may be responsible for the form of the decay of light-curves of supernovae of Type I which have an exponential form with a half-life of 55 nights. They described the way in which ^{254}Cf may be synthesized in a supernovea outburst, and reasons why the energy released by its decay may do-

5.37   M. Taketani, T. Hatanaka, and S. Obi, Progress of Theoret. Phys. 15:89 (1956)
5.38   F. Hoyle and M. Schwarzschild, Astrophys. J. Supplement II, No. 13 (1955)
5.39   A.E. Roy, Mon. Not. R. A. S. 112:484 (1952)
5.40   R.J. Tayler, Astrophys. J. 120:332 (1954)
5.41   M. Schwarzschild and L. Spitzer, Observatory 73:77 (1953)
5.42   W. Baade, G.R. Burbidge, F. Hoyle, E.M. Burbidge, R.F. Christy and W.A. Fowler, Pub. Astron. Soc. Pacific 68 No. 403, 296 (1956)
5.43   G.R. Burbidge, F. Hoyle, E.M. Burbidge, R.F. Christy and W.A. Fowler, Phys. Rev. 103:1145 (1956)

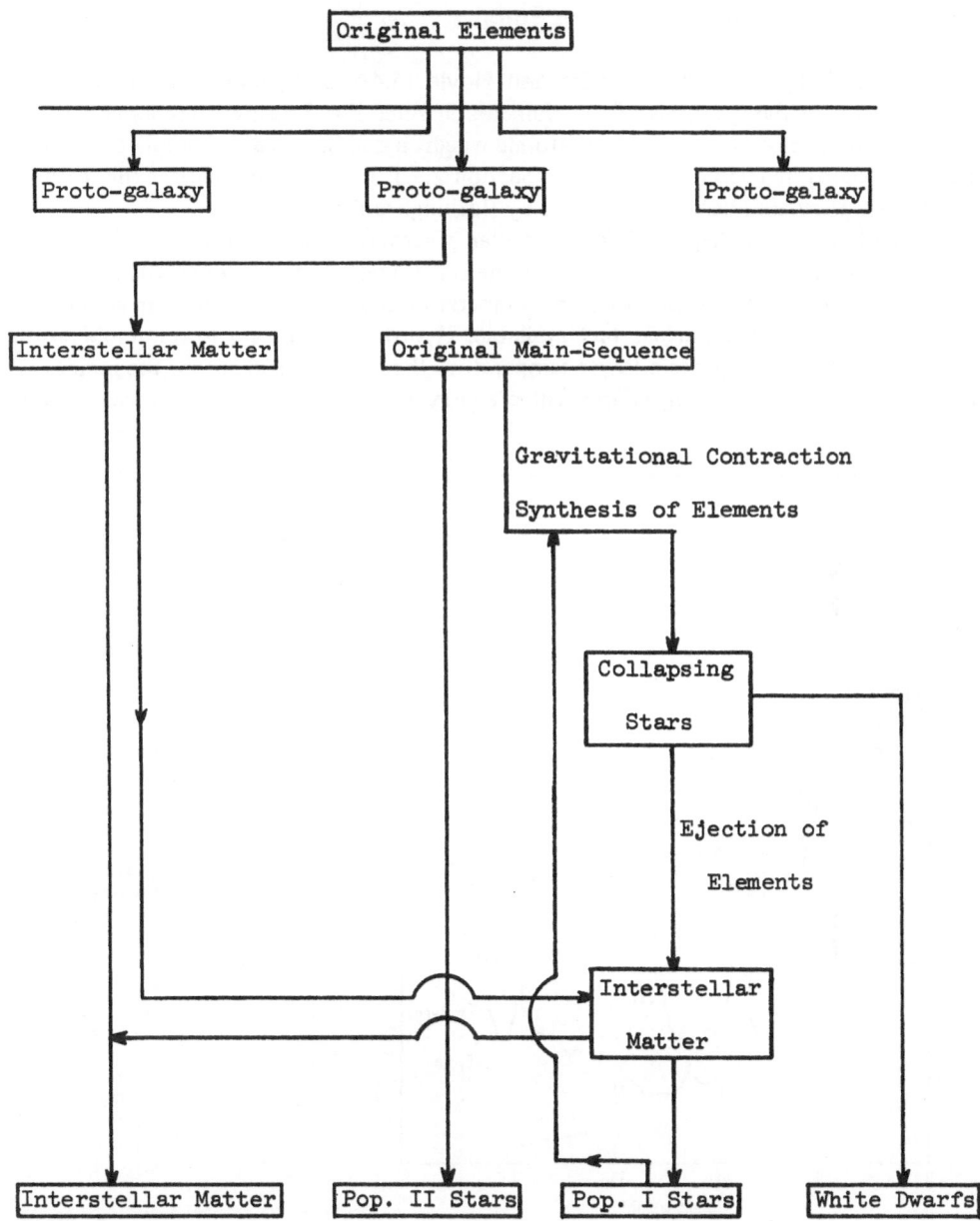

Fig. 5.2. The evolutionary scheme of stars proposed by Taketani, Hatanaka and Obi (Progress of Theoretical Physics 15:89 (1956)

minate all others. They also noted that the presence of Tc in red giant stars and of Cf in Type I supernovae appears to be observational evidence that neutron capture processes on both a slow and a fast timescale have been necessary to synthesize the heavy elements in their observed cosmic abundances.

5.12. Synthesis of the Elements in Stars

In 1957, Burbidge, Burbidge, Fowler and Hoyle [5.44, 5.45] published the so-called B_2FH theory for the synthesis of the elements in stars. Fig. 5.3 shows a schematic curve of atomic abundances as a function of atomic weight based on the data compiled by Suess and Urey [5.46] in 1956 (see Chapter 2, Section 2.9.). Table 5.2 summarizes the features of the Suess-Urey abundance curve noted by Burbidge and co-workers.

According to the theory of B_2FH, a star possesses a self-governing mechanism in which the temperature is adjusted so that the outflow of energy through the star is balanced by nuclear energy generation. The temperature required to give this adjustment depends on the particular nuclear fuel available. Hydrogen requires a lower temperature than helium; helium requires a lower temperature than carbon and so on, the increasing temperature sequence ending at iron since energy generation by fusing processes ends

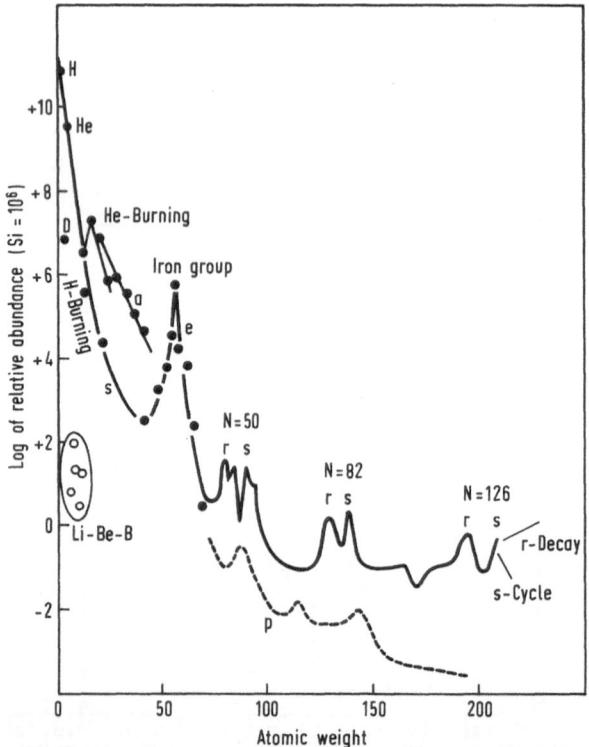

Fig. 5.3. Schematic curve of atomic abundances as a function of atomic weight based on the data compiled by Suess and Urey. There is still considerable spread of the individual abundances about the curve illustrated, but the general features shown are now fairly well established. Note the overabundance relative to their neighbors of the alpha-particle nuclei A = 16, 20, --- 40, the peak at the iron group nuclei, and the twin peaks at A = 80 and 90, at 130 and 138, and at 194 and 208: B_2FH, Rev. Mod. Phys. 29:547 (1957)

5.44   E.M. Burbidge, G.R. Burbidge, W.A. Fowler and F. Hoyle, Rev. Mod. Phys. 29:547 (1957)
5.45   E.M. Burbidge and G.R. Burbidge, Science 128:387 (1958)
5.46   H.E. Suess and H.C. Urey, Rev. Mod. Phys. 28:53 (1956)

Table 5.2. Features of the Suess-Urey abundance curve, according to B_2FH

Feature	Cause
Exponential decrease from hydrogen to A ~100	Increasing rarity of synthesis for increasing A, reflecting that stellar evolution to advanced stages necessary to build high A is not common.
Fairly abrupt change to small slope for A > 100	Constant σ (n, γ) in s process. Cycling in r process.
Rarity of D, Li, Be, B as compared with their neighbors H, He, C, N, O	Inefficient production, also consumed in stellar interiors even at relatively low temperatures.
High abundances of alpha-particle nuclei such as O^{16}, Ne^{20} ... Ca^{40}, Ti^{48} relative to their neighbors	He burning and α process more productive than H burning and s process in this region.
Strongly-marked peak in abundance curve centered on Fe^{56}	e process; stellar evolution to advanced stage where maximum energy is released (Fe^{56} lies near minimum of packing-fraction curve).
Double peaks $\{$ A = 80, 130, 196	Neutron capture in r process (magic N = 50, 82, 126 for progenitors).
A = 90, 138, 208	Neutron capture in s process (magic N = 50, 82, 126 for stable nuclei).
Rarity of proton-rich heavy nuclei	Not produced in main line of r or s process; produced in rare p process.

here. If hydrogen is present the temperature is adjusted to hydrogen as a fuel, and is comparatively low. If hydrogen becomes exhausted as stellar evolution proceeds, the temperature rises until helium becomes effective as a fuel. The automatic temperature rise is brought about in each case by the conversion of gravitational energy into thermal energy.

In order to explain all the features of the Suess-Urey abundance curve shown in Fig. 5.3 and Table 5.2, at least eight different types of synthesizing processes are demanded, if it is assumed that only hydrogen is primordial.

I. Hydrogen Burning. This process includes the so-called proton-proton chain (5.13) and the carbon-nitrogen cycle (5.12), and also the nuclear reactions which synthesize the isotopes of carbon, nitrogen, oxygen, fluorine, neon and sodium which are not produced by the processes II. and III.

II. Helium Burning. These processes are responsible for the synthesis of carbon from helium, and by further α-particle addition for the production of ^{16}O, ^{20}Ne, and perhaps ^{24}Mg.

III. α-Process. These processes include the reactions in which α particles are successively added to ^{20}Ne to synthesize the four structure nuclei ^{24}Mg, ^{28}Si, ^{32}S, ^{36}A, ^{40}Ca, and probably ^{44}Ca and ^{48}Ti. The source of the α particles is different in the α process than in helium burning.

IV. e-Process. This is the so-called equilibrium process previously discussed by Hoyle [5.47] in which under conditions of very high temperature and density the elements comprising the iron peak in the Suess-Urey abundance curve (vanadium, chromium, manganese, iron, cobalt, and nickel) are synthesized.

V. s-Process. This is the process of neutron capture with the emission of gamma radiation (n, γ) which takes place on a long time scale, ranging from about 10^2 to about 10^5 years for each neutron capture. The neutron captures occur at a *slow* (s) rate compared to the intervening beta decays. This mode of synthesis is responsible for the production of the majority of the isotopes in the range $23 \leqslant A \leqslant 46$ (excluding those synthesized predominantly by the α process), and for a considerable proportion of the isotopes in the range $63 \leqslant A \leqslant 209$. The s process produces the abundance peaks at A = 90, 138, and 208.

VI. r-Process. This is the process of neutron capture on a very short time-scale, about $0.01 - 10$ sec for the beta-processes interspersed between the neutron captures. The neutron captures occur at a *rapid* (r) rate compared to the beta decays. This mode of synthesis is responsible for production of a large number of isotopes in the range $70 \leqslant A \leqslant 209$, and also for synthesis of uranium and thorium. This process may also be responsible for some light element synthesis, e.g., ^{36}S, ^{46}Ca, ^{48}Ca, and perhaps ^{47}Ti, ^{49}Ti, and ^{50}Ti. The r-process produces the abundance peaks at A = 80, 130, and 194.

VII. p-Process. This is the process of proton capture with the emission of gamma radiation (p, γ), or the emission of a neutron following gamma-ray absorption (γ, n), which is responsible for the synthesis of a number of proton-rich isotopes having low abundances as compared with the nearby normal and neutron-rich isotopes.

VIII. x-Process. This process is responsible for the synthesis of deuterium, lithium, beryllium, and boron. More than one type of process may be demanded here. The characteristic of all of these elements is that they are very unstable at the temperatures of stellar interiors, so that it appears probable that they have been produced in regions of low density and temperature (see Chapter 7, Section 7.1, Origin of Lithium, Beryllium and Boron).

A schematic diagram of the nuclear processes by which the synthesis of the elements in stars takes place is shown in Fig. 5.4.

5.13. The e-Process According to B$_2$FH

At temperatures above approximately $3 \cdot 10^9$ degrees all manner of nuclear processes occur in great profusion: i.e., (γ, α), (γ, p), (γ, n), (α, γ), (p, γ), (n, γ), (p, n) reactions as well as others involving heavier nuclei. Hoyle [5.47] reported that the abundances of the elements in the iron peak could be synthesized under conditions of temperature and density such that statistical equilibrium between the nuclei and the free protons and neutrons

---

5.47   F. Hoyle, Month. Not. Roy. Astron. Soc. 106:343 (1946); Astrophys. J. Suppl. I:121 (1954)

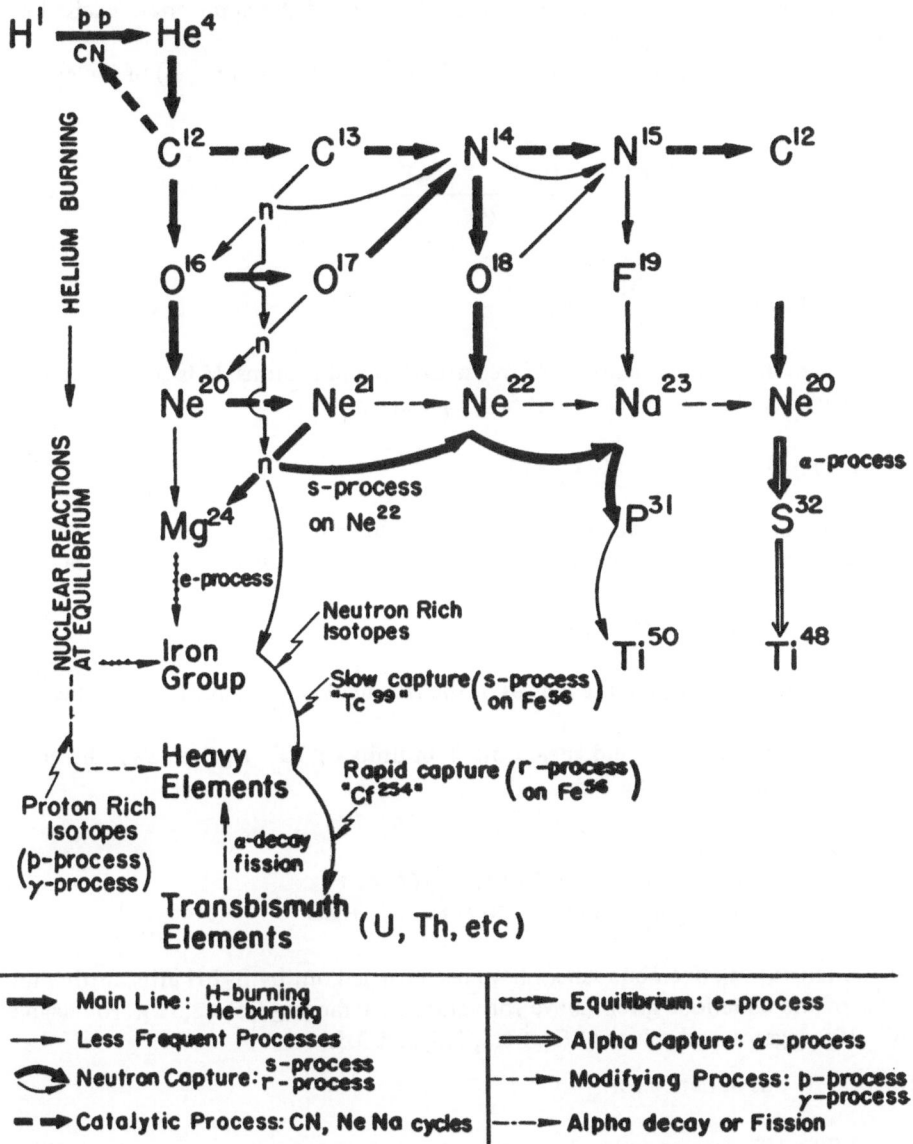

Fig. 5.4. A schematic diagram of the nuclear processes by which the synthesis of the elements in stars takes place. Elements synthesized by interactions with protons (hydrogen burning) are listed horizontally. Elements synthesized by interactions with alpha particles (helium burning) and by still more complicated processes are listed vertically. The details of the production of all of the known stable isotopes of carbon, nitrogen, oxygen, fluorine, neon, and sodium are shown completely. Neutron capture processes by which the highly charged heavy elements are synthesized are indicated by curved arrows. The production of radioactive Tc^{99} is indicated as an example for which there is astrophysical evidence of neutron captures at a slow rate over long periods of time in red giant stars. Similarly, Cf^{254}, produced in supernovae, is an example of neutron synthesis at a rapid rate. The iron group is produced by a variety of nuclear reactions at equilibrium in the last stage of a star's evolution: B_2FH, Rev. Mod. Phys. 29:547 (1957)

was achieved. B_2 FH, in 1957, re-examined the *e*-process using the more accurate nuclear and abundance data then available, and taking into account the appropriate excited nuclear states, the energies and spins.

Under conditions of statistical equilibrium the number density n(A, Z) of the nucleus A, Z is given by

$$n(A,Z) = \omega(A,Z) \; \frac{A \, M \, kT^{3/2}}{2 \, \pi \, \hbar^2} \cdot \frac{(n_n)^{A-Z} \, (n_p)^Z}{2A} \times$$

$$\left(\frac{2 \, \pi \, \hbar^2}{M \, kT} \right)^{3A/2} \cdot \exp \frac{Q(A,Z)}{kT}$$

(5.25)

where n_n, n_p are the number densities of free neutrons and protons, M is the atomic mass unit, and $\omega(A, Z)$ is the statistical weight factor given by

$$\omega(A, Z) = \sum_r (2I_r + 1) \exp(-E_r/kT)$$

(5.26)

where E_r is the energy of the excited state measured above the ground level, and I_r is the spin. $Q(A, Z)$ is the binding energy of the ground level of the nucleus A, Z, and is given by

$$Q(A, Z) = c^2 \, [(A - Z) \, M_n + ZM_p - M(A, Z)]$$

(5.27)

where M_n, M_p, and M (A, Z) are the masses of the free neutron, free proton, and nucleus A, Z, respectively.

Writing $\Theta = \log(n_p/n_n)$, and measuring T in units of 10^9 degrees (T_9), the above equation can be expressed in the form

$$\log n(A,Z) = \log \omega(A,Z) + 33.77 +$$
$$(3/2) \log(AT_9) + 5.04/T \cdot Q(A,Z) +$$
$$A(\log n_n - 34.07 - (3/2) \log T_9) + Z\Theta$$

(5.28)

Calculations of relative abundances have been carried out by B_2 FH after further modifications of the equations given above for values of T ranging from 2.52×10^9 degrees to 7.56×10^9 degrees, and values of Θ of 1.5, 2.5, and 3.5.

The best fit was obtained for

$$T = 3.78 \cdot 10^9 \text{ degrees and}$$
$$\Theta = 2.5$$

(5.29)

This value of Θ corresponds to a density of the order of 10^5 g/cc.

5.14. The *s*- and *r*-Processes According to B_2 FH

In the buildup of nuclei by the *s* and *r* processes the reactions which govern both the rate of flow and the track followed in the (A, Z) plane are the (n, γ) and (γ, n) reactions, beta decay, and at the ends of the tracks, alpha decay in the case of the *s* process and neutron-induced fission in the case of the r process. According to B_2FH, the rate of the (n, γ), (γ, n) and beta process are

$$
\begin{aligned}
\gamma_n &= 1/\tau_n = \sigma_n \, v_n \, n_n, \\
\lambda_\beta &= 1/\tau_\beta = \text{const}/W_\beta^5, \\
\lambda_\gamma &= 1/\tau_\gamma = \sigma_\gamma \, c \, n_\gamma
\end{aligned}
\tag{5.30}
$$

where σ_n and σ_γ are the cross sections for the (n, γ) and (γ, n) reactions, respectively; v_n and n_n are the velocity and density of neutrons responsible for the (n, γ) reactions; n_γ is the density of γ radiation; and W_β is the beta-decay energy.

The general equation for the buildup of nuclei in the s-process is then

$$
\begin{aligned}
dn(A,Z)/dt = {}&\lambda_n \, (A{-}1,Z) \, n(A{-}1,Z) - \lambda_n \, (A,Z) \, n \, (A,Z) + \\
&\lambda_\beta \, (A,Z{-}1) \, n \, (A,Z{-}1) - \lambda_\beta \, (A,Z) \, n \, (A,Z) + \\
&\text{termination terms due to alpha decay at } A > 209
\end{aligned}
\tag{5.31}
$$

The general equation for the r-process is

$$
\begin{aligned}
dn(A,Z)/dt = {}&\lambda_n \, (A{-}1,Z) \, n \, (A{-}1,Z) - \lambda_n \, (A,Z) \, n \, (A,Z) + \\
&\lambda_\beta \, (A,Z{-}1) \, n \, (A,Z{-}1) - \lambda_\beta \, (A,Z) \, n \, (A,Z) + \\
&\lambda_\gamma \, (A{+}1,Z) \, n \, (A{+}1,Z) - \lambda_\gamma \, (A,Z) \, n \, (A,Z) + \\
&\text{termination terms due to fission for } A \geqslant 260
\end{aligned}
\tag{5.32}
$$

For the s process, we have, in general,

$$
\lambda_n < \lambda_\beta \, (\tau_n > \tau_\beta)
\tag{5.33}
$$

For the r process, we have

$$
\lambda_n > \lambda_\beta \, (\tau_n < \tau_\beta)
\tag{5.34}
$$

As long as $\lambda_n > \lambda_\beta$, buildup continues with Z constant and λ_n continuously decreases until

$$
\lambda_n \, (A,Z) \approx \lambda_\gamma \, (A{+}1,Z)
$$

At this point no further buildup can take place until beta decay occurs, thereby increasing Z. The effective rate of neutron addition at this point is such that

$$
\lambda_n \, (A,Z) - \lambda_\gamma \, (A{+}1,Z) \; < \lambda_\beta \, (A,Z)
\tag{5.35}
$$

Fig. 5.5 shows the neutron capture paths of the s and r processes obtained by B_2FH. A detailed mathematical treatment of the creation of heavy elements by the r process was also reported by Fong [5.48].

According to B_2FH, of the eight processes which are demanded to synthesize all of the stable isotopes, assignments among hydrogen burning, helium burning, the α process,

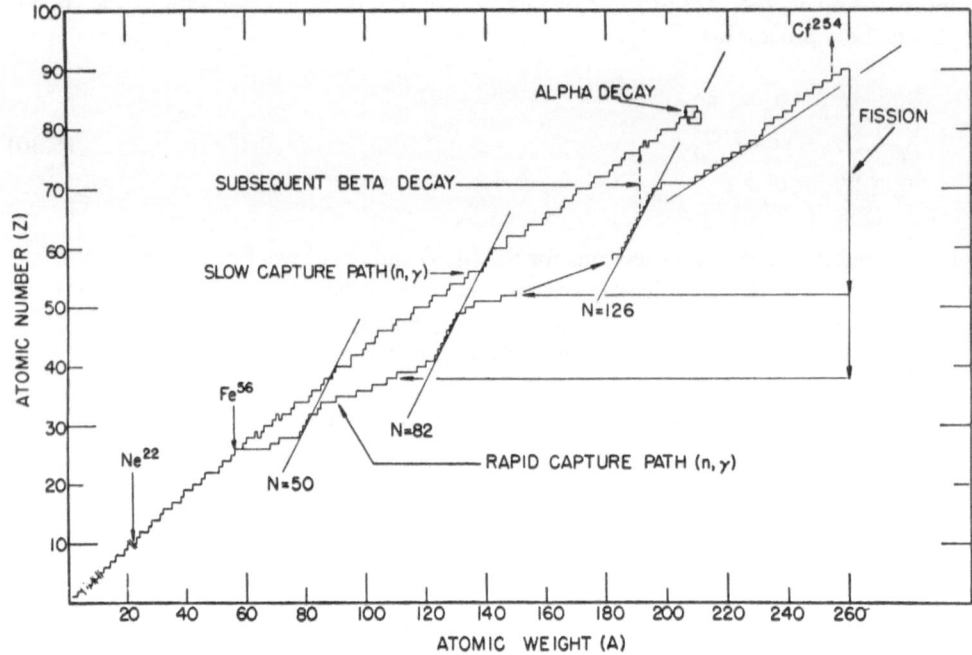

Fig. 5.5. The neutron capture paths of the *s* and *r* processes, according to B₂FH, Rev. Mod. Phys. 29: 547 (1957)

the *e* process, and the *x* process are comparatively straightforward. As far as assignments among the *s, r,* and *p* processes are concerned, the situation is a little more complex. Suess and Urey [5.46] and Coryell [5.49] pointed out that the peaks in the abundance curves at stable nuclei with filled neutron shells

$$A = \ 90, N = \ 50;$$
$$A = 138, N = \ 82; \quad\quad\quad (5.36)$$
$$A = 208, N = 126$$

strongly indicate the operation of the *r* process, and the nearby peaks at A = 80, 130, and 194, shifted by δ A ∼8 to 14, similarly require the operation of the *r* process.

Calculations of Fowler et al. [5.50] also suggested that apart from the nuclei built by the α process all of the nuclei with 23 ⩽ A ⩽ 46 with the exception of ^{36}S, ^{46}Ca, and ^{48}Ca can be synthesized by the *s* process. Certain isotopes of the heavy elements can be built only by the *r* process while others can be built only by the *s* process. In some cases, both processes contribute to the synthesis of one isotope.

Fig. 5.6 shows how the isotopes in the mass 120—150 region were assigned to the *s, r,* and *p* processes.

5.48   P. Fong, Phys. Rev. 120:1388 (1960)
5.49   C.D. Coryell, J. Chem. Ed. 38:67 (1961)
5.50   W.A. Fowler, G.R. Burbidge and E.M. Burbidge, Astrophys. J. 122:271 (1955)

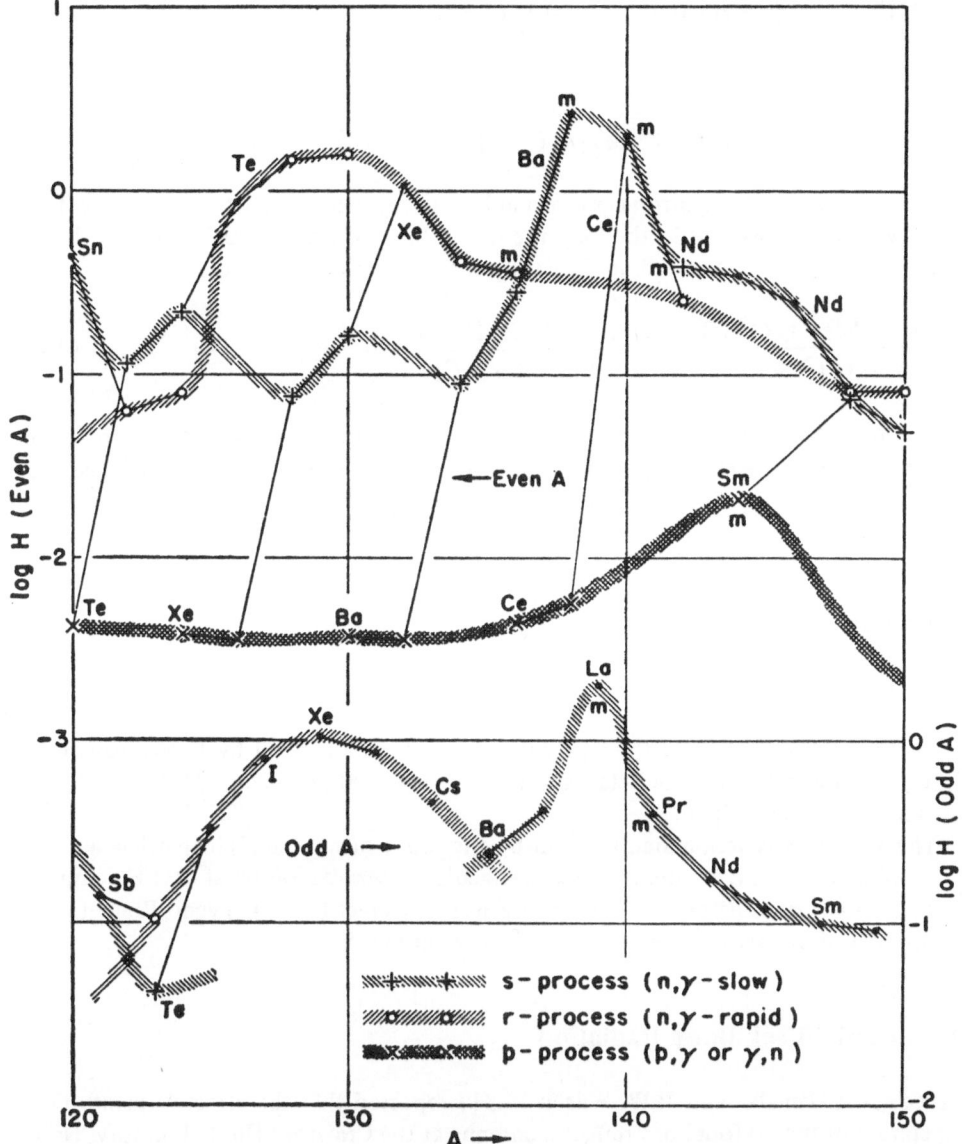

Fig. 5.6. Assignment of isotopes in the mass 120–150 region to the *s*, *r*, and *p* processes, according to B₂FH, Rev. Mod. Phys. 29:547 (1957). +: produced only in the *s* process, ∘: only in the *r* process, •: produced in both processes, x: only in the *p* process

According to B_2FH, there are two indicators for the time scale for the *s* process in two regions of A. The first can be obtained from two isotopes of krypton. Both ^{80}Kr and ^{82}Kr are built only by the *s* process, being shielded in the *r* process by ^{80}Se and ^{82}Se. But, there is a "loop" at ^{79}Se, because the beta-decay half-life of this isotope is $7 \cdot 10^4$ years. Thus, while ^{82}Kr is produced partly in the main *s* process chain and partly in the loop, ^{80}Kr is only produced in the loop. Now under conditions of steady neutron flow, the basic equations for the relative abundances of these isotopes are

77

$$N\,(^{80}\mathrm{Kr})\,\lambda_n\,(^{80}\mathrm{Kr}) = 0.92N\,(^{79}\mathrm{Se})\,\lambda_\beta\,(^{79}\mathrm{Se}) \qquad (5.37)$$

and

$$N\,(^{82}\mathrm{Kr})\,\lambda_n\,(^{82}\mathrm{Kr}) = N\,(^{79}\mathrm{Se})\,[\lambda_\beta\,(^{79}\mathrm{Se}) + \lambda_n\,(^{79}\mathrm{Se})] \qquad (5.38)$$

where λ_n and λ_β are the neutron-capture and beta-decay rates, respectively. The coefficient 0.92 appears because in the loop the nuclei have to pass through ^{80}Br which beta decays only 92 percent of the time. From the above equations, we have

$$\frac{t_n\,(^{79}\mathrm{Se}) + t_\beta\,(^{79}\mathrm{Se})}{t_n\,(^{79}\mathrm{Se})} = 0.92\,\frac{t_n\,(^{80}\mathrm{Kr})\,N\,(^{82}\mathrm{Kr})}{t_n\,(^{82}\mathrm{Kr})\,N\,(^{80}\mathrm{Kr})} \qquad (5.39)$$

The neutron capture cross section for ^{80}Kr is about twice that for ^{82}Kr and hence $\lambda_n\,(^{80}\mathrm{Kr}) \sim 2\,\lambda_n\,(^{82}\mathrm{Kr})$ and $t_n\,(^{82}\mathrm{Kr}) \sim 2\,t_n\,(^{80}\mathrm{Kr})$. From the known abundance ratio $N(^{82}\mathrm{Kr})/N(^{80}\mathrm{Kr})$, we have from the above equation

$$t_n/t = 1/1.37$$

and, since $t = 7 \cdot 10^4$ years,

$$t_n = 5.1 \cdot 10^4 \text{ years} \qquad (5.40)$$

The second indicator for the s process time-scale pointed out by B_2FH was the region of ^{152}Gd and ^{154}Gd. The beta half-life of ^{151}Sm is 90 years and ^{152}Gd is mostly bypassed in the s-process chain.

The s, process is terminated at ^{209}Bi by the onset of natural alpha-particle activity. The termination of the r process is the spectacular supernova outburst. B_2FH compared the light curves of supernovae observed by Baade, Kepler (1604), Tycho Brahe (1572) and also by Chinese astronomers (1054), as shown in Fig. 5.7.

5.15. Cosmic Black-Body Radiation

In 1965, A.A. Penzias and R.W. Wilson [5.51] reported that the effective zenith noise temperature of the 20-foot horn-reflector antenna at the Crawford Hill Laboratory, Holmdel, New Jersey, at 4080 Mc/s yielded a value about 3.5 °K higher than expected. This excess temperature was, within the limits of their observations, isotropic, unpolarized, and free from seasonal variations (July, 1964 – April, 1965). They stated that a possible explanation for the observed excess noise temperature was the one given by Dicke et al. [5.52]: radiation coming from space at radio and far infrared wavelengths, believed to be energy left over from the Big-Bang origin of the Universe.

5.51   A.A. Penzias and R.W. Wilson, Astrophys. J. 142:419 (1965)
5.52   R.H. Dicke, P.J.E. Peebles, P.G. Roll, and D.T. Wilkinson, Astrophys. J. 142:414 (1965)

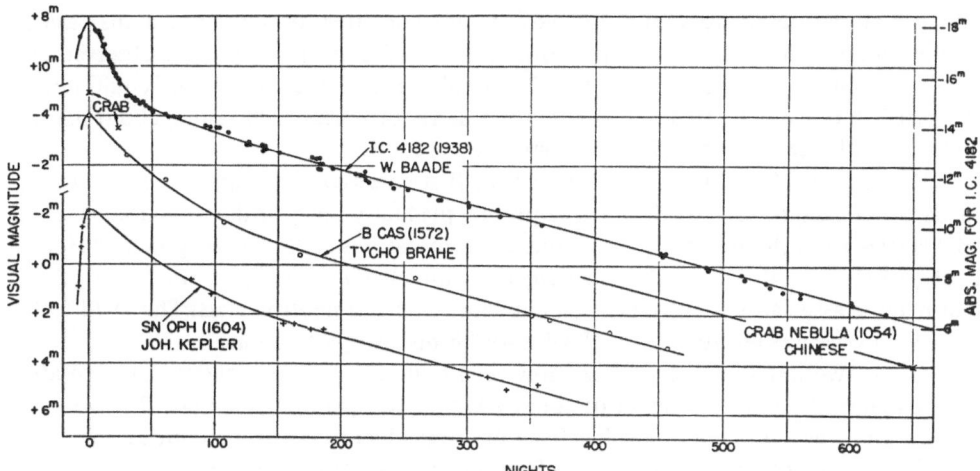

Fig. 5.7. The termination of the *r* process, according to B$_2$FH: light curves of supernovae. Measurements for SN IC 4182 are by Baade; those for B Cassiopeiae (1572) and SN Ophiuchi (1604) have been converted by Baade to the modern magnitude scale from the measures by Tycho Brahe and Kepler. The three points for the supernova of 1054 are uncertain, being taken from the ancient Chinese records (N.U. Mayall and J.H. Oort, Publ. Astron. Soc. Pacific 54:95 (1942)). The abscissa gives the number of nights after maximum; the left-hand ordinate gives the apparent magnitude (separate scale for each curve); the points for the Crab Nebula belong on the middle scale. The right-hand ordinate gives the absolute magnitude for SN IC 4182 derived by using the current distance scale: B$_2$FH, Rev. Mod. Phys. 29:547 (1957)

In 1967, Gamow [5.53] reported that the theory of the expanding universe, first proposed by A. Friedman [5.54] in 1922, has recently received an encouraging boost from the observational disapproval of the steady-state cosmology and the unexpected detection of the residual isotropic thermal 3 °K radiation by Penzias and Wilson, the existence of which was predicted on the basis of evolutionary cosmology almost two decades ago [5.55].

The 1978 Nobel Prize in Physics was awarded to A.A. Penzias and R.W. Wilson for their discovery of cosmic black-body radiation. The Big-Bang theory of Gamow used to share the limelight with the Steady State theory, which holds that the universe had no beginning and is eternal, but the discovery of the cosmic fireball by Penzias and Wilson provided clenching evidence that the universe really did explode into existence.

5.16. Pulsars or Neutron Stars

In 1968, A. Hewish and co-workers [5.56] recorded unusual signals from pulsating radio sources at the Mullard Radio Astronomy Observatory, Cavendish Laboratory, University of Cambridge. A large radio telescope operating at a frequency of 81.5 MHz was brought

5.53 G. Gamow, Science 158:766 (1967)
5.54 A. Friedmann, Z. Phys. 10:377 (1922)
5.55 G. Gamow, Phys. Rev. 74:505 (1948); Nature 162:680 (1948)
5.56 A. Hewish, S.J. Bell, J.D.H. Pilkington, P.F. Scott and R.A. Collins, Nature 217:709 (1968)

into use at the observatory in July 1967. Soon thereafter, it was noticed that signals which appeared at first to be weak sporadic interference were repeatedly observed at a fixed declination and right ascension. This result showed that the source could not be terrestrial in origin. The signals consisted of a series of pulses each lasting about 0.3 seconds and with a repetition period of about 1.337 seconds.

The remarkable nature of these signals at first suggested an origin in terms of man-made transmissions, such as deep space probes, planetary radar or the reflexion of terrestrial signals from the moon. None of these interpretations could be accepted and Hewish et al. [5.56] stated that the radiation seemed to come from local objects within the galaxy, and may be associated with oscillations of white dwarf or neutron stars. The 1974 Nobel Price in Physics was awarded to A. Hewish for his discovery of pulsars.

About two hundred pulsars have now been found. Pulsars are believed to be rotating neutron stars. They have a radius of about 10 kilometers, although in mass they are comparable with the sun. One pulsar has been positively identified with an optical telescope. It is located in the very center of the Crab nebula. Neutron stars are formed as a final stage in the evolution of supernovae, and it is likely that supernovae also give rise to black holes. A theoretical object whose gravitational pull is so strong that nothing can escape, not even light, is called a black hole. The idea was first suggested by the French mathematician Pierre Laplace in 1795, on the basis of Newton's theory of gravity. A black hole with a mass equal to that of the sun is about 6 kilometers across.

5.17. The World of Antimatter

In 1969, H. Alfvén and A. Elvius [5.57] proposed a new theory of the evolution of galaxies: the "symmetric cosmology" leading to a theory of quasars (quasi-stellar objects). A quasar is an object which appears as a star-like point of light, but which emits far more energy than an entire galaxy. Quasars were discovered in 1963 by Maarten Schmidt and over 200 quasars are now known. In his Nobel Lecture, Alfvén [5.58] also discussed the problems related to plasma physics, space research and the origin of the solar system. The symmetric approach to cosmology — without the assumption of any *ad hoc* laws of nature — leads necessarily to the conclusion that there should be antimatter in our galaxy as well as in all galaxy. Presumably one half of the mass of a galaxy should consist of ordinary matter and the other half, of antimatter. The symmetric approach of Alfvén and Elvius appears to be irreconcilable with the "Big-Bang" theory of Gamow, because a homogenous mixture of matter and antimatter having the density required by the big-bang theory would result in immediate annihilation. The universe described by the advocates of the big-bang theory is "asymmetric" in the sense that it contains ordinary matter but no antimatter.

5.57   H. Alfvén and A. Elvius, Science 164:911 (1969)
5.58   H. Alfvén, Nobel Lecture, 11 December 1970; Science 172:991 (1971)

5.18. Possible Climatic Effect of Supernova Explosions

A number of investigators have postulated that nearby supernova explosions may have affected terrestrial life and could have caused the extinction of certain exposed animals during the past 600 million years of the earth's history [5.59, 5.60]. It is not a simple matter, however, to test these ideas experimentally, because the occurrence of a nearby supernova explosion is an extremely rare event. Supernova explosions within 50 light-years, for example, occur at intervals of a few hundred million years or less.

Since cosmic rays originate primarily from supernovae, and since they produce ^{14}C in the earth's atmosphere, the effect of supernova explosions should be detectable in the form of an enhanced ^{14}C level at the surface of the earth. According to Lingenfelter and Ramaty [5.61], the discovery of pulsars, which are neutron-star remnants of supernovae, possibly provides new clues to the age and distance of supernova remnants. They suggested that the largest cosmic-ray flux at present would be from PSR1929+10 with an age of about 10^5 years and is located at a distance of 260 light-years. The maximum flux at the earth from this source would have arrived about 9×10^4 years ago and the flux would now be decreasing by the $-3/2$ power of the age. Thus it is possible that long-time variations in the rate of ^{14}C production in the earth's atmosphere may reflect changes in the local cosmic-ray flux resulting from nearby supernova explosions.

The supernovae observed in historical times are all so distant that cosmic rays from them have not yet reached the earth. It appears to be plausible, however, that x-rays and gamma-rays from relatively distant supernova explosions may have a measurable climatic effect. Kuroda [5.62] has recently pointed out the possibility that gamma-rays from relatively distant supernova explosions (for example, the Crab Nebula, which is located at a distance of about 3,600 to 5,500 light years) may cause an increase of the nitrogen oxide inventory of the atmosphere, which in turn may lead to a reduction of O_3 and a world-wide change in climate. Such an effect may be detectable in the temperature data obtained by the method of isotopic tree thermometers invented by L.M. Libby and co-workers [5.63].

5.19. Search for Neutrinos from the Sun

The sun produces energy by means of a series of fusion reactions and the overall process can be represented as the synthesis of ^{4}He nuclei from four hydrogen nuclei by the reaction shown by equation (5.11). The positrons and gamma radiation are rapidly absorbed in the material of the sun and their energy is converted into heat. Neutrinos, however, are neutral particles with zero mass and have unique ability to pass through hundreds of light-years of ordinary matter without being absorbed. Davis and co-workers [5.64] have

5.59   K.D. Terry and W.H. Tucker, Science 159:421 (1968)
5.60   M.A. Ruderman, Science 184:1079 (1974)
5.61   R.E. Lingenfelter and R. Ramaty, Nobel Symposium 12, Radiocarbon Variations and Absolute Chronology, edited by I.U. Olsson, John Wiley, 1970, p. 513
5.62   P.K. Kuroda, Geochem. J. 11:45 (1977)
5.63   L.M. Libby, L.J. Pandolfi, P.H. Payton, J. Marshall III, B. Becker and V. Giertz-Siebenlist, Nature 261:284 (1976)
5.64   R. Davis, Jr., D.S. Harmer and K.C. Hoffman, Phys. Rev. Letters 20:1205 (1968)

attempted to observe this neutrino radiation from the sun. Such an experiment, if successful, would enable scientists to test by direct observation the thermonuclear origin of the solar energy. Investigators would be able to check the detailed reaction mechanisms by comparing the measured neutrino fluxes with those calculated from theoretical models of the sun.

A radiochemical solar neutrino detector was built by Davis et al. [5.64] of the Brookhaven National Laboratory. It depends upon the capture of neutrinos in ^{37}Cl to produce radioactive ^{37}Ar (35-day half-life) by the reaction

$$\nu + {}^{37}\text{Cl} \rightarrow {}^{37}\text{Ar} + e^- \tag{5.41}$$

A very large volume, 390,000 liters, of tetrachloroethylene (C_2Cl_4) was used as a neutrino-capturing medium. ^{37}Ar atoms produced in the neutrino detector were recovered quantitatively, purified and placed in a low-level counter to observe the characteristic radiations from the decay of 35-day ^{37}Ar. The neutrino detector was built 4850 ft underground in the Homestake gold mine, Lead, S. Dakota.

The solar neutrino detector built by Davis and co-workers has a high sensitivity for neutrinos above 6.0 MeV energy and hence it is primarily sensitive to neutrinos from ^{8}B decay in the sun, which is produced in the following side reactions of the proton-proton chain:

$$
\begin{aligned}
{}^3\text{He} + {}^4\text{He} &\rightarrow {}^7\text{Be}, \\
{}^7\text{Be} + {}^1\text{H} &\rightarrow {}^8\text{B}, \\
{}^8\text{B} &\rightarrow {}^8\text{Be} + e^+ + \nu \\
{}^8\text{Be} &\rightarrow 2\,{}^4\text{He}.
\end{aligned}
$$

In 1968, Davis et al. [5.64] reported that a search was made for solar neutrinos and the upper limit of the product of the neutrino flux and the cross sections for all sources of neutrinos was 3×10^{-36} sec^{-1} per ^{37}Cl atom. They concluded that the flux of neutrinos from ^{8}B decay in the sun was equal to or less than 2×10^6 cm^{-2} sec^{-1} at the earth, and that less than 9 percent of the sun's energy appeared to be produced by the carbon-nitrogen cycle (see also many papers published on the subject of solar neutrinos and the experiment of Davis and co-workers [5.65–5.72]).

5.65   R. Davis, Jr., Phys. Rev. Letters 12:303 (1964)
5.66   J. Bahcall, Phys. Rev. Letters 12:300 (1964)
5.67   J.N. Bahcall, Science 147:115 (1965)
5.68   R. Davis, Jr., McGraw-Hill Yearbook of Science and Technology, 1969
5.69   W.A. Fowler, Nuclear Astrophysics, American Philosophical Society, Independence Square, Philadelphia, 1967, p. 46
5.70   G. Shaviv and E.E. Salpeter, Phys. Rev. Letters 21, 1602 (1968)
5.71   V. Linke, Naturwissenschaften 58:77 (1971)
5.72   F. Hoyle, Astronomy and Cosmology: A Modern Course, W.H. Freeman and Company, San Francisco, 1975; p. 387

5.20. Temperature of the Sun

In 1971, the writer [5.73] reported that he attempted to determine the temperature of the sun shortly after formation, from the difference in the isotopic compositions of the xenon found in the earth's atmosphere and on the surface of the moon. He noted that the latter was transported from the sun in the form of the solar wind and had excesses of ^{128}Xe, ^{130}Xe, ^{131}Xe and ^{132}Xe relative to the terrestrial xenon, which could be explained as due to the neutron-capture reactions which occurred in the core of the sun during its deuterium-burning stage. The ratio of the excess ^{130}Xe and the excess ^{132}Xe found in the lunar soil appeared to be very nearly 1, which is not the ratio expected from the capture at room temperature. The ^{132}Xe/^{130}Xe ratio of the excess xenon at these mass numbers may thus provide information about the ratio of the cross-section values for neutron capture at a temperature much higher than room temperature. If so, we may be able to use this ratio as a "thermometer" for the measurement of the temperature of the environment where the capture reactions occurred.

If such a thermometer is to be developed for determining the temperature of the sun's interior, the cross-section values for neutron-capture covering the neutron energy range 100 ev to 10 KeV are required for the xenon isotopes. Unfortunately, data are not available at this time, and we can only pursue this idea on the basis of estimated cross-section values. The writer showed that such a rough estimate leads to a value of about 8×10^6 K, a value not far from the theoretical reaction temperature for the ^{2}D(d, n)^{3}He reaction which is 1.2×10^6 K (see Section 5.4., Table 5.1).

5.21. Further Studies on Nucleosynthesis in Stars

In 1975, Virginia Trimble [5.74] published a paper entitled "The Origin and Abundances of the Chemical Elements" and reviewed the results obtained from the studies on nucleosynthesis in stars prior to April 1, 1975. In this excellent review article, she discussed a number of important subjects in great detail. For example, she noted that some doubt has been cast on the entire scheme of hydrogen burning in stars (see Section 5.5) by the failure of R. Davis (see Section 5.19) to detect any neutrinos coming form the Sun, and added that the apparent absence of solar neutrinos was a serious problem, which may yet require us to rethink our entire picture of stellar energy generation and nucleosynthesis.

She also presented a detailed discussion on the subject of isotopic anomalies observed in oxygen, neon, and xenon, and also on the subject of nucleocosmochronology. It is interesting to note here that she concluded the discussion by stating: "Finally, it should be said that, within the conventional picture, we do not understand the origins of Ne-E, the excess ^{16}O, or the anomalous, fission-like Xe, or the sites at which the ^{16}O, iodine, and xenon are retained". These problems related to the so-called *xenology* and the isotopic anomalies in the early solar system will be fully discussed in Chapters 6 and 7 of this monograph. The fact that no one process seems capable of producing all the "deficient" light nuclides, ^{2}H, ^{3}He, ^{4}He, ^{6}Li, ^{9}Be, ^{10}B, and ^{11}B was also pointed out by her. It is also worthy of note that she pointed out that Suess and Zeh [5.75] have presented a dis-

5.73   P.K. Kuroda, Nature Physical Science 230:40 (1971)
5.74   Virginia Trimble, Rev. Mod. Phys. 47:877 (1975)
5.75   H.E. Suess and H.D. Zeh, Astrophys. Space Sci. 23:123 (1973)

cussion of the abundances of the heavy nuclides and a critique of the standard approach to nucleosynthesis, while Amiet and Zeh [5.76, 5.77] suggested an alternative way of forming the neutron-rich heavy nuclei.

It should be also noted here that Bethe's C-N-O cycle (see Section 5.5) seems to supply a minor part of the solar energy production, mainly based on the proton-proton reactions according to equation (5.11), whereas it plays a perceptible role in stars considerably heavier than the Sun [5.78]. Lambert and Ries [5.79] have reported that C_2 and CN molecular spectra in certain red giant stars showed anomalous $^{12}C : {}^{13}C : {}^{14}N$ ratios, indicating the importance of Bethe's C-N-O catalyzed hydrogen-helium fusion in these special cases.

5.76   J.P. Amiet and H.D. Zeh, Phys. Lett. B25:305 (1967)
5.77   J.P. Amiet and H.D. Zeh, Z. Phys. 217:676 (1968)
5.78   B. Kuchowicz, Reports on Progress in Physics 39:291–343 (1976)
5.79   D.L. Lambert and L.M. Ries, Astrophys. J. 248:228 (1981)

6.
Plutonium-244 in the Early Solar System

"If elements heavier than uranium exist it is probable that they will be radioactive. The extreme delicacy of radioactivity as a means of chemical analysis would enable such elements to be recognized even if present in infinitesimal quantity."

E. Rutherford and F. Soddy, Phil. Mag (6) 5:576 (1903)

Plutonium-244 (half-life, 8.2×10^7 years) and iodine-129 (half-life, 1.7×10^7 years) existed in nature during the early history of the solar system. The occurrence of plutonium-244 in the early solar system can be considered as proof that the transuranium elements were synthesized by the *r*-process in exploding stars (supernovae). There remained, however, uncertainties concerning the origin of iodine-129 and the interpretation of the isotopic compositions of xenon in meteorites.

6.1. Rutherford and Soddy's View on the Transuranium Elements

In 1903, E. Rutherford and F. Soddy [6.1] published an important paper, in which they put forward the law of exponential radioactive decay. In the same paper, they speculated on the possible existence of the elements heavier than uranium and stated: "If elements heavier than uranium exist it is probable that they will be radioactive. The extreme delicacy of radioactivity as a means of chemical analysis would enable such elements to be recognized even if present in infinitesimal quantity. It is therefore to be expected that the number of radio-elements will be augmented in the future, and that considerably more than *three* at present recognized exist in minute quantity."

The three elements which they referred to in 1903 were radium, thorium and uranium. Each had a series of disintegration products, but Rutherford and Soddy felt that these were not real elements. It seemed to be advisable to give a special name for these

---

6. 1   E. Rutherford and F. Soddy, Phil. Mag. (6) 5:576 (1903)

atom-fragments, or new atoms, which result from the original atom after the ray had been expelled, and which remain in existence only a limited time, continually undergoing further change. Rutherford and Soddy suggested the term *metabolon* for these atom fragments. It thus appears that Rutherford and Soddy, in 1903, were thinking of the possible existence of elements heavier than uranium in *nature* — elements originally created in stars.

6.2. Rutherford's Calculation of the Age of the Elements

In 1929, Rutherford [6.2] published a short note in *Nature* in which he stated: "By the kindness of Dr. Aston, I have had the opportunity of inspecting his photographs showing the isotopes of lead obtained from the radioactive mineral bröggerite. As he concludes, it seems highly probable that the isotope of mass 207 is mainly due to actinium lead, and that the actinium series has its origin in an isotope of uranium — a suggestion independently put forward by several investigators on other evidence. In the light of this new knowledge and of the measurements made by Dr. Aston of the relative intensities of the lead isotopes in the mineral, it may be of interest to consider its bearings on the origin of actinium and other problems."

Rutherford called this new isotope "actino-uranium" and guessed that its mass to be 235, although he considered the less likely case of an isotope with $A = 239$ and $Z = 92$. He estimated the half-life of actino-uranium as follows:

If λ_1 and λ_2 are the decay constants of actino-uranium and ^{238}U, respectively,

$$K'/K = \frac{\lambda_2}{\lambda_1} \cdot \frac{e^{\lambda_1 t} - 1}{e^{\lambda_2 t} - 1} \tag{6.1}$$

where t is the age of the mineral from which the lead is derived; K' is the ratio of the number of atoms of actinium lead to those of uranium lead, which could be deduced from Aston's measurements: K is the ratio of the number of atoms delivered in a mineral into the actinium series compared with the number passing into the radium series.

Rutherford assumed that t to be about 10^9 years — an average estimate of the age of old primary uranium minerals. K was known to be about 3/100 in the 1920's. Taking as a low estimate that $K' = 7/100$, it could be deduced from the above equation that $\lambda_1/\lambda_2 = 10.6$.

The half-life of ^{238}U was then known to be 4.5×10^9 years and hence the half-life of ^{235}U turned out to be 4.2×10^8 years, which is considerably shorter than the modern value of 7.1×10^8 years. Taking the period as 4.2×10^8 years, Rutherford calculated the amount of actino-uranium remaining on the earth today to be only about 0.28 percent of the main uranium isotope — an amount too small to influence appreciably the atomic weight of uranium as ordinarily measured.

If it is supposed that uranium, like other heavy elements, was formed from stellar matter, it is likely that actino-uranium of odd atomic weight would be formed in smaller quantity than the main isotope of even atomic weight. Even, however, if we suppose they

---

6. 2   E. Rutherford, Nature 123:313 (1929)

were formed in equal quantity, it can be shown that it would require only 3.4×10^9 years to bring down the amount to the 0.28 percent observed today. Rutherford thus felt that the earth could not be much older than 3.4×10^9 years, if we suppose that the production of uranium in the earth ceased as soon as the earth separated from the sun. Moreover, if the age of the sun is of the order of magnitude estimated by Jeans (7×10^{12} years), it appeared that the uranium isotopes which we observe in the earth must have been forming in the sun at a late period of its history, namely, about 4×10^9 years. Rutherford then made the following interesting statements: "We may thus conclude, I think with some confidence, that the processes of production of elements like uranium were certainly taking place in the sun 4×10^9 years ago and probably still continue today." To update this statement, one needs to change the word sun to supernovae and increase the number 4 to about 5.

6.3. The Concept of Extinct Radioactivity

It has long been known to geologists that small colored areas exist in certain kinds of mica, for example, in biotite, cordierite, and muscovite. They are called pleochroic haloes and the centers of these areas usually contain a minute crystal of foreign matter. Joly [6.3], in 1907, first pointed out that these haloes are of radioactive origin and they are due to the coloration of mica by the α rays expelled from a nucleus which contains radioactive matter. In the complete uranium halo, the outer ring marks the range of the α particles from radium C' (7 cm in air). In the thorium halo it corresponds to the range of the α particles from thorium C' (8.6 cm in air).

Joly later observed in some specimens of mica very small X haloes and Iimori and Yoshimura [6.4] found Z haloes whose radii were not in agreement with known ranges of the uranium and thorium series. Rutherford, Chadwick and Ellis [6.5] remarked in 1930: "While it is very difficult to be certain of the exact origin of these minute haloes, a systematic examination of varied materials is required to see whether any definite evidence can be obtained of the existence of additional types of radioactive matter in the early history of the earth. The subject is one of great interest and importance, as the study of haloes gives us a method of detection of possible radioactive changes in matter which does not survive in detectable amounts today."

In 1947, Brown [6.6] published a short note entitled "An Experimental Method for the Estimation of the Age of the Elements," in which he considered the time elapsed since the formation of the elements in two stages:

t_1 = the time interval between the formation of the elements and the separation of elements into the various phases comprising meteoritic matter; and

t_2 = the time interval between the meteoritic phase separation and the present.

The short-lived and intermediate-lived isotopes must have since decayed away, while those species with half-lives comparable to the total elapsed time ($t_1 + t_2$) still exist in nature.

6. 3   J. Joly, Phil. Mag. 13:381 (1907)

6. 4   S. Iimori and J. Yoshimura, Inst. Phys. Chem. Res. Tokyo 5:11 (1926)

6. 5   E. Rutherford, J. Chadwick and C.D. Ellis, Radiations from Radioactive Substances, Cambridge at the University Press, 1951 (Re-issue of the 1930 edition), p. 183

6. 6   H. Brown, Phys. Rev. 72:348 (1947)

Now suppose, (a) if the meteorite phase separation took place before a given radio-active species had decayed into insignificance, and (b) if the parent and daughter elements were fractionated from one another during the phase separation, then one would expect the isotopic composition of the daughter element to differ in the two phases. In such a case, one would be able to set an upper limit to the time interval of a few half-lives.

In 1948, Aten [6.7] noted that ^{129}Xe is more than four times as abundant as its two neighboring isotopes together and offered the following explanation for this interesting phenomenon: after the concentration of the original xenon in the earth's atmosphere had fallen to its present low value, the ^{129}Xe formed by the beta-decay of the ^{129}I, which at that time had a much higher concentration in the iodine of the earth's outer crust and of the ocean, mixed with the small amount of the original xenon.

In 1948, Suess [6.8] pointed out that the knowledge of the exact half-life of ^{129}I would be of great interest, because it would enable one to determine the time between the synthesis of the elements and the formation of the earth's atmosphere and predicted that the half-life of ^{129}I must be at least by a factor of 10 smaller than the time interval between the element synthesis and formation of the earth's atmosphere. The abundance of iodine on the earth's surface is about 1 gram per cm^2. At the time of the element synthesis, the abundance of the long-lived ^{129}I must have been about the same order of magnitude as that of the stable ^{127}I. The abundance of ^{129}Xe from the decay of ^{129}I in the atmosphere, however, appears to be only about 10^{-4} gram per cm^2. Suppose that the ratio ^{129}I : ^{127}I by the time of formation of the earth's atmosphere was 10^{-3}, then the abundance of the ^{129}Xe should have been abnormally high, just as in the case of ^{40}Ar in the atmosphere. The abundance of ^{129}Xe in the earth's atmosphere, however, is not higher by a factor of 2 than expected from the normal pattern of abundance systematics.

Suess, in 1948, also pointed out that the ^{129}Xe anomaly originates from the abnor-mal distribution of the total isobaric abundances for the masses 128 and 130. In other words, ^{128}Xe and ^{130}Xe appear to be much less abundant and the ^{128}Te and ^{130}Te correspondingly much more abundant.

6.4. Half-Life of Iodine-129 and the Age of the Elements

In 1951, Katcoff and co-workers [6.9] at the Brookhaven National Laboratory measured the half-life of ^{128}I to be 17.2 ± 0.9 million years and reported that the time interval between the formation of the elements and the formation of the earth to be 2.7×10^8 years. The half-life of ^{129}I was determined by measuring the absolute disintegration rates, isotopic compositions, and total iodine contents of several samples obtained from fission product iodine that was separated from a uranium slug which had received a 4-years irra-diation in the Oak Ridge pile and had cooled for 21 months.

Katcoff et al. [6.9] calculated the time interval (t) between the formation of the ele-ments and the formation of the earth's atmosphere assuming that the atmosphere contains both the radiogenic ^{40}Ar from the decay of ^{40}K and the radiogenic ^{129}Xe from the decay of ^{129}I which once existed in the earth's crust. They wrote:

6. 7   A.H.W. Aten, Jr., Phys. Rev. 73:1206 (1948)

6. 8   H.E. Suess, Z. f. Physik 125:386 (1948)

6. 9   S. Katcoff, A.O. Schaeffer and J.M. Hastings, Phys. Rev. 82, 688 (1951)

$$({}^{127}\text{I} \cdot e^{-\lambda t}/{}^{40}\text{K})_T = ({}^{129}\text{Xe}/{}^{40}\text{Ar}) \qquad (6.2)$$

where ${}^{40}\text{Ar}$ refers to its abundance in the atmosphere by volume (0.93%); ${}^{129}\text{Xe}$ refers to the radiogenic ${}^{129}\text{Xe}$ which was *assumed* to be 20 percent of the total Xe abundance in the atmosphere, 8×10^{-6} percent Xe by volume: λ is the decay constant of ${}^{129}\text{I}$; and the subscript T refers to the ratio for the earth.

Katcoff et al. [6.9] felt that this left 6.23 percent ${}^{129}\text{Xe}$ which is not radiogenic, and hence the rule of Aten is approximately satisfied, and the rule which states that two odd isotopes of an element have roughly the same abundance (the rule of Suess and Mattauch) is not seriously violated either, since the abundance of ${}^{131}\text{Xe}$ is 21.2 percent. Solving the above equation, they obtained a value of t = 15.4 half-lives of ${}^{129}\text{I}$, or 2.7×10^8 years.

Soon after Katcoff et al.'s paper was published, Suess and Brown [6.10] pointed out that it may not be correct to assume that the bulk of ${}^{129}\text{Xe}$ in the earth's atmosphere was produced by the decay of ${}^{129}\text{I}$. The reason for this is that the abundance of ${}^{129}\text{Xe}$ is approximately the same as that of ${}^{131}\text{Xe}$. Thus, it is more likely that the abundances of ${}^{128}\text{Xe}$ and ${}^{130}\text{Xe}$ are actually abnormally *low*, rather than the abundance of ${}^{129}\text{Xe}$ being abnormally high. The reason for the exceptionally low abundance of ${}^{128}\text{Xe}$ and ${}^{130}\text{Xe}$ is found by considering the systematics of cosmic abundance of nuclides: it appears that the two abundant tellurium isotopes (${}^{52}\text{Te}^{128}$ and ${}^{52}\text{Te}^{130}$) *shield* the two xenon isotopes. This leads to the apparently high abundance of ${}^{129}\text{Xe}$, which happens to be *unshielded*. For this reason, Suess and Brown suggested that the value of t is more likely to be 4×10^8 years, rather than 2.7×10^8 years.

6.5. Excess ${}^{129}\text{Xe}$ in Meteorites

In 1960, Reynolds [6.11] made the important discovery that the xenon from the chondritic (stone) meteorite Richardton is heavily enriched in ${}^{129}\text{Xe}$. He concluded that this isotope almost certainly was formed from the radioactive decay of ${}^{129}\text{I}$, now extinct as a natural radioactivity but not so at the time of formation of the meteorites, and he calculated that $(0.35 \pm 0.06) \times 10^9$ years elapsed between the time of formation of the elements and the meteorite.

Since Brown suggested in 1947 that the meteorites could be used to determine quite accurately the age of the elements if the daughter of an extinct natural radioactivity could be found there, the decay

$${}^{129}\text{I} \xrightarrow{1.7 \times 10^7 \text{ yr}} {}^{129}\text{Xe} \text{ (stable)}$$

has long been recognized as a particularly favorable case for stone meteorites, although previous searches for fossil ${}^{129}\text{Xe}$ in the chondrite Beardsley by Wasserburg and Hayden [6.12] and the achondrite Nuevo Laredo by Reynolds and Lipson [6.13] failed to give a positive result.

6.10   H.E. Suess and H. Brown, Phys. Rev. 83:1254 (1951)
6.11   J.H. Reynolds, Phys. Rev. Letters 4:8 (1960)
6.12   G.J. Wasserburg and R. Hayden, Nature 176:130 (1955)
6.13   J.H. Reynolds and J. Lipson, Geochim. Cosmochim. Acta 12:330 (1957)

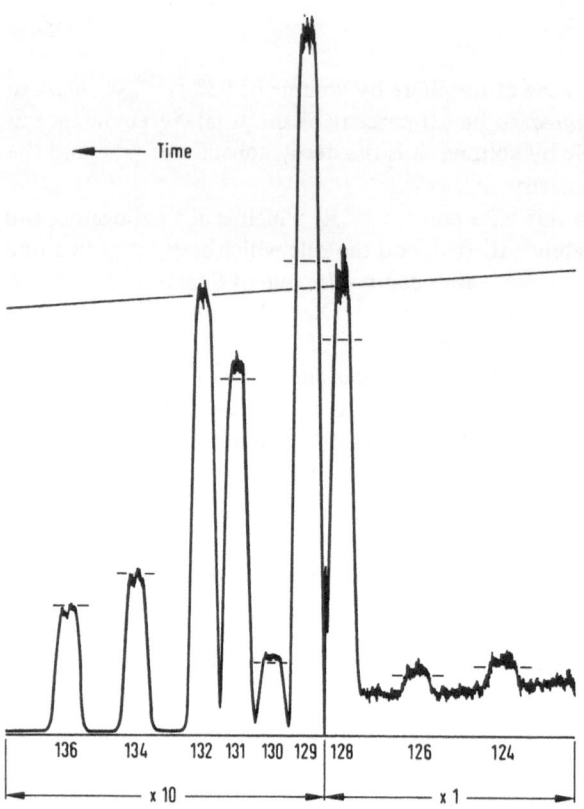

Fig. 6.1. Mass spectrum of xenon extracted from a 7-gram sample of Richardton meteorite (J.H. Reynolds, Phys. Rev. Letters 4:8, 1960). Horizontal lines show the comparison spectrum of terrestrial Xe. The slant line through the 132 peak shows the extent of spectrometer pumping. Jagged peak tops are due to statistical fluctuations in the small ion currents.

Fig. 6.1 shows the mass spectrum of xenon extracted from a 7-gram sample of Richardton meteorite. The horizontal lines drawn through the various peaks show where the peaks would fall for an analysis, under identical conditions, of a sample of atmospheric xenon [6.14] having the same ^{132}Xe content. Reynolds [6.11] assumed that the ^{136}Xe in the sample of xenon from the meteorite was an index to the xenon of terrestrial composition in the sample, whether by contamination from the atmosphere or otherwise. The isotopic composition of the xenon from the meteorite, after this terrestrial component has been subtracted, is shown in Table 6.1.

Reynolds [6.11] considered various other possible mechanisms for ^{129}Xe production, as shown in Table 6.2. No other process produces excesses of ^{129}Xe and ^{131}Xe in such a manner that the ^{129}Xe/^{131}Xe ratio is as high as 10.0. He thus concluded that ^{129}I decay has produced the bulk of the ^{129}Xe. It can be calculated from Table 6.1 that one gram of the Richardton meteorite contains 13×10^{-11} cc STP of "excess" ^{129}Xe. This amount was much larger than the upper limit of 1.3×10^{-11} cc STP/g for the Beardsley meteorite reported by Wasserburg and Hayden [6.12].

6.14   A.O. Nier, Phys. Rev. 79:450 (1950)

Table 6.1. Xenon from 6.69 g of Richardton meteorite. Isotopic composition of anomalous component (Reynolds, 1960)

Isotope	124	126	128	129	130	131	132	134	136
Percent abundance	0.33 ±0.09	0.23 ±0.02	2.5 ±0.3	81.5 ±3.5	2.3 ±0.4	8.2 ±2.0	3.6 ±2.3	1.4 ±1.0	≡0

Xenon of terrestrial composition: $(0.88 \pm 0.05) \cdot 10^{-9}$ cc STP/g
Xenon of anomalous composition: $(0.16 \pm 0.01) \cdot 10^{-9}$ cc STP/g

Table 6.2. Possible mechanisms for ^{129}Xe production considered by Reynolds

Mechanism	$(^{129}Xe/^{131}Xe)$
Slow-particle or spontaneous fission	< 1
Fast-particle fission [6.15]	> 1 but ~1
Spallation reactions on Cs, Ba, or heavier elements	< 1 but ~1
Natural neutron capture in Te [6.16]	3
Observed	10.0

The time interval between nucleosynthesis and the formation of the meteorite can be calculated by the equation:

$$\Delta t = \frac{(1.72 \times 10^7)}{0.693} \left[\ln \left(^{127}I/^{129}Xe\right)_{met} + \ln \left(^{129}I/^{127}I\right)_0 \right] \text{ years} \tag{6.3}$$

It was assumed that $(^{129}I/^{127}I)_0 = 1$, i.e., that ^{129}I and ^{127}I were produced initially in equal amounts and an estimated iodine content of the meteorite of 1 ppm was used in the calculation. This leads to

$$\ln \left(^{127}I/^{129}Xe\right)_{met} = 14.1 \tag{6.4}$$

and $\Delta t = 0.35 \times 10^9$ years. Note that in these calculations, an error of a factor of 10 in either $(^{127}I)_{met}$ or $(^{129}I/^{127}I)_0$ leads to only a 16 percent error in Δt.

One curious feature of the isotopic composition of the anomalous xenon component was noted. This was the fact that, excluding ^{129}Xe, the abundances of the 5 lightest isotopes fitted a smooth curve when normalized to the abundances of atmospheric xenon. In Fig. 6.2 these normalized abundances have been plotted logarithmically against $\sqrt{M}$. This raised the possibility that the anomalies at masses 124—131 may not be due to nuclear

6.15   P.C. Stevenson et al., Phys. Rev. 111:886 (1958)
6.16   M. Inghram and J.H. Reynolds, Phys. Rev. 78:822 (1950)

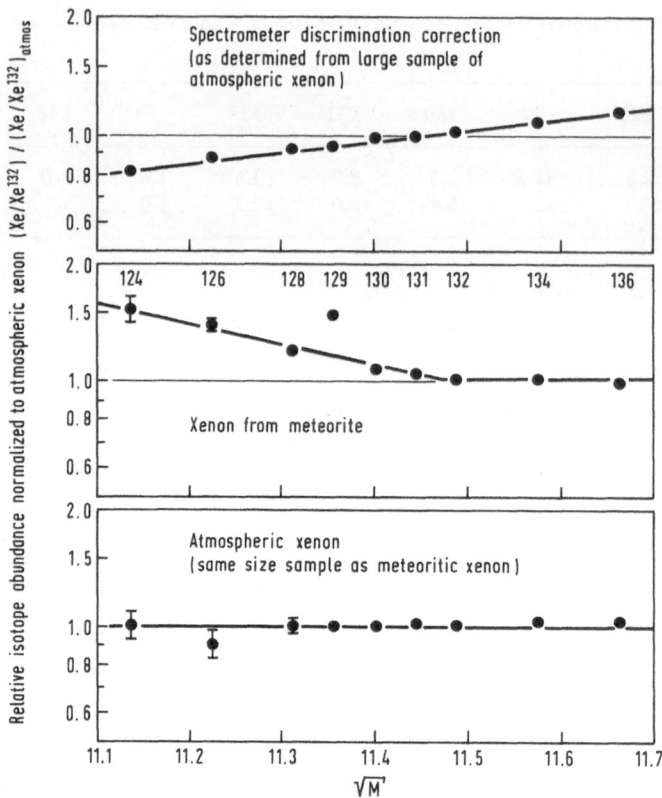

Fig. 6.2. Relative isotope abundance of meteoritic xenon (Richardton) normalized to atmospheric xenon, according to J.H. Reynolds, Phys. Rev. Letters 4:8 (1960)

processes at all, but may be due to a strong mass fractionation of meteoritic xenon relative to terrestrial xenon — such as might occur if terrestrial xenon is the final residue of gas which has mostly escaped from a gravitational field at some time during its history.

The heavy isotopes 132, 134, and 136 did not fit the pattern and Reynolds [6.11] noted that it was probably significant that these were the chief isotopes produced in slow particle and spontaneous fission. He stated that *the meteoritic spectrum could be decomposed quite well into a "fractionated" terrestrial component (including masses, 132, 134, and 136) plus a fission xenon spectrum.* This puzzling feature of the isotopic composition of xenon remained not clearly understood for many years and led to the idea concerning the existence of the so-called carbonaceous chondrite fission (CCF) xenon in meteorites (see Section 6.9., Unsolved Problems in Xenology and Chapter 7, Isotopic Anomalies in the Early Solar System, Section 7.7., Xenon).

Shortly after the paper of Reynolds was published, Wasserburg, Fowler and Hoyle [6.17] pointed out that Reynold's Δt value is the maximum time interval which could exist between the termination of nucleosynthetic processes and the formation of the Richardton meteorite and it does not date the time of formation of the elements. It was

---

6.17   G.J. Wasserburg, W.A. Fowler and F. Hoyle, Phys. Rev. Letters 4:112 (1960)

assumed by Reynolds [6.11] that the nucleosynthetic processes during which iodine was made were instantaneous, in accord with such short-time scale models as proposed by Alpher, Bethe and Gamow [6.18]. On the other hand, Burbidge, Burbidge, Fowler and Hoyle [6.19] proposed that nucleosynthesis took place in stars at a more or less uniform rate over the lifetime of the Galaxy and the period of nucleosynthesis was taken by these investigators to be of the order of 10^{10} years.

If the production of iodine occurs uniformly over a period of time T, the following relationship should hold:

$$\ln ({}^{127}I/{}^{129}Xe) = \Delta t/\tau + \ln (T/\tau) + \ln (K_{127}/K_{129}) \qquad (6.5)$$

where K_{127} and K_{129} are the production rates of ^{127}I and ^{129}I, respectively, and τ is the mean life of ^{129}I (2.5×10^7 years).

Reynolds' experiment yields

$$\ln ({}^{127}I/{}^{129}Xe) = 14.1$$

and in interpreting the experimental result, Reynolds took $T < \tau$, in which case $\ln (T/\tau)$ does not occur in the equation. Then assuming

$$K_{127}/K_{129} \sim 1$$

Reynolds found $\Delta t = 3.5 \times 10^8$ years.

Wasserburg et al. [6.17] reported that it was quite likely that $T \gg \tau$, and the most reasonable values which fit the data of Reynolds are $T = 10^{10}$ years and $\Delta t = 2 \times 10^8$ years.

6.6. The Plutonium-244 Hypothesis

Shortly after the paper by S. Katcoff and co-workers [6.9] was published in 1951, the writer felt that the disappearance of ^{129}I during the early history of the earth might explain the fact that the atomic weight of iodine (126.9) is smaller than that of tellurium (127.6). Such a process, however, would produce an effect of decreasing the atomic weight of xenon (131.30) by enriching the relative abundance of ^{129}Xe. In order to counterbalance such an effect, it appeared to be necessary that xenon became "heavier", because of the addition of the heavy isotopes $^{131-136}Xe$ produced by the process of fission:

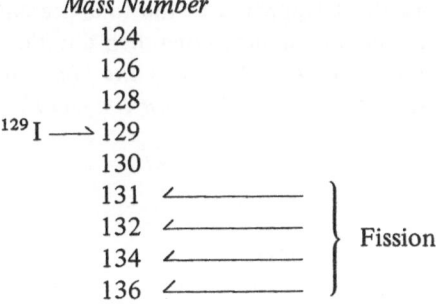

6.18   R.A. Alpher, H.A. Bethe and G. Gamow, Phys. Rev. 73:803 (1948)
6.19   E.M. Burbidge, G.R. Burbidge, W.A. Fowler and F. Hoyle, Rev. Mod. Phys. 29:547 (1957)

There appeared to be two possibilities: (a) self-sustaining uranium chain reactions could have occurred on the earth and/or (b) some unknown transuranium elements could have decayed by spontaneous fission during the early history of the earth. These possibilities were discussed by the writer in two papers published in 1956 [6.20] and 1960 [6.21]. The idea (a) led to the investigations on the Oklo Phenomenon, which was discussed in Chapter 4. In regard to the possibility (b) that some transuranium elements might have decayed by spontaneous fission during the early history of the earth, Seaborg and co-workers [6.22], in 1951, were aware of the fact that one of the yet-to-be discovered isotopes of element 94, ^{244}Pu, may have a half-life as long as 100 million years, and if so, it may still remain on the earth.

In 1951, Peppard and co-workers [6.23] isolated microgram quantities of plutonium from a uranium process waste from the processing of Belgian Congo pitchblende corresponding to an order of magnitude of 100 tons of ore. A mass spectrographic analysis of a sample containing one microgram of ^{239}Pu showed, however, that no other isotope of plutonium was present in a concentration as great as one percent of the ^{239}Pu. Peppard and co-workers [6.24] also isolated a small quantity of ^{237}Np, the long-lived ancestor of the (4n + 1) radioactive series from a uranium process waste.

In 1954, M.H. Studier and co-workers [6.25] at Argonne National Laboratory discovered the long-lived isotope of plutonium, ^{244}Pu, in the debris from the November 1952 thermonuclear test conducted in the South Pacific. The half-life of the newly-discovered ^{244}Pu turned out to be 82 million years or 1/56 of the age of the earth, which is 4.6 billion years. It thus became apparent by the middle of the 1950's that the primordial ^{244}Pu had decayed to an almost infinitesimal fraction of its initial abundance and hence it would be futile to attempt to detect it on the earth today.

In 1957, Aten [6.26] published a paper entitled "The Calculation of the Age of the Elements," in which he distinguished four phases in the formation of transuranic nuclei which today exist as uranium and thorium. In the first stage nuclei were formed many of which had an appreciable neutron-excess. Aten called these the "original" nuclei. During the succeeding period these nuclei changed into β-stable isobar by means of β-decay. The quantities which existed after β-stabilization are called "stabilized" quantities. Now the "fast" α-processes set in, which were much slower than the β-decays, but much faster than the "slow" α-decay of ^{238}U, ^{235}U and ^{232}Th. During this period the decay of ^{238}U, ^{235}U and ^{232}Th has been negligible. Some of these fast α-processes were followed by even faster β-decays and the final products formed were ^{238}U, ^{235}U and ^{232}Th.

Both the first and the second period were very short compared to the total present age of the elements and it appears as if we are fully justified in neglecting their duration. At this point, however, Aten added the following remark: *The slowest of the "fast" α-decays, that of ^{244}Pu, might have placed a small role if it had not been for the fact that*

6.20   P.K. Kuroda, J. Chem. Phys. 25:781 (1956)
6.21   P.K. Kuroda, Nature 187:36 (1960)
6.22   C.A. Levine and G.T. Seaborg, J. Am. Chem. Soc. 73:3278 (1951)
6.23   D.F. Peppard, M.H. Studier, M.V. Gergel, G.W. Mason, J.C. Sullivan and J.F. Mech, J. Am. Chem. Soc. 73:2529 (1951)
6.24   D.F. Peppard, G.W. Mason, P.R. Gray and J.F. Mech, J. Am. Chem. Soc. 74:6081 (1952)
6.25   M.H. Studier, P.R. Fields, P.H. Sellers, A.M. Friedman, C.M. Stevens, J.F. Mech, H. Diamond, J. Sedlet and J.R. Huizenga, Phys. Rev. 93:1433 (1954)
6.26   A.H.W. Aten, Jr., Physica 23:1073 (1957)

only a relatively small fraction of the present ^{232} *Th has come from or passed through this nuclide.*

It appeared to the writer that what Aten said in 1957 meant that a significant amount of ^{244}Pu must have existed in nature shortly after the earth was formed and, if so, it may have partially decayed by spontaneous fission to produce the heavy xenon isotopes in the earth's atmosphere as shown in Fig. 6.3.

To test this idea, it was only necessary to compare the isotopic composition of the xenon in the earth's atmosphere with that of the xenon in extraterrestrial samples such as meteorites. The rare gas mass-spectrometer, which enables one to perform such an experiment, was just being constructed at about this time by Reynolds [6.27] at the Department of Physics of the University of California at Berkeley. It was three years later in 1960 that Reynolds [6.11] made the important discovery that the xenon extracted from the meteorite Richardton was highly enriched in ^{129}Xe and he attributed this enrichment to the decay of 17 million-year ^{129}I during the early history of the solar system.

If 17 million-year ^{129}I existed in the early solar system, 82 million-year ^{244}Pu should have also been present in nature, and the xenon extracted from the meteorite should have been enriched in the heavy isotopes of xenon. When the isotopic composition of xenon in Richardton meteorite was compared with that of the atmospheric xenon, however, it was found that exactly the opposite was the case: ^{129}Xe was enriched, but $^{131-136}$Xe were depleted in the meteorite relative to the xenon in the atmosphere. Thus the result obtained by Reynolds appeared to indicate that ^{244}Pu did not exist in nature in the early solar system. This would have been too hasty a conclusion, however.

In 1960, Kuroda [6.21] presented the following arguments: (a) the difference between the isotopic compositions of the xenon in the meteorite Richardton and in the atmosphere could be interpreted as due to the existence of excess fission xenon in the latter; (b) the difference indicated that at least 10 percent of the atmospheric ^{136}Xe was fissiogenic and this was much greater than that expected from the ^{238}U spontaneous fission alone, but (c) the difference could be explained as due to the spontaneous fission of some of the "extinct" transuranium elements and/or the induced fission of ^{235}U in the early history of the earth. He mentioned the ^{244}Pu as the prime candidate for the production of excess fission xenon.

It so happened that the above arguments (a) and (b) were grossly in error. As it will be discussed in Chapter 7, Isotopic Anomalies in the Early Solar System, Section 7.7., Xenon, the difference between the isotopic compositions of the xenon in most of the meteorites and in the atmosphere can be explained as due to a combined effect of mass-fractionation, neutron capture and cosmic-ray irradiation processes which occurred in the early solar system and the contribution from the excess fission xenon in the atmosphere amounts to roughly one percent, instead of 10 percent. The argument (c), on the other hand, turned out to be not totally in error. As it will be discussed in Section 6.8., Plutonium-244 in the Early Solar System, it turned out that the difference in the isotopic compositions of the xenon in the meteorite Pasamonte and in the atmosphere could be interpreted as due to the existence of excess fission xenon from ^{244}Pu in the former.

Although most of the calculations which were carried out by the writer in 1960 turned out to be not quite correct, the essence of the arguments presented at that time may not be totally useless and hence will be briefly outlined below.

6.27   J.H. Reynolds, Rev. Sci. Instr. 27:928 (1956)

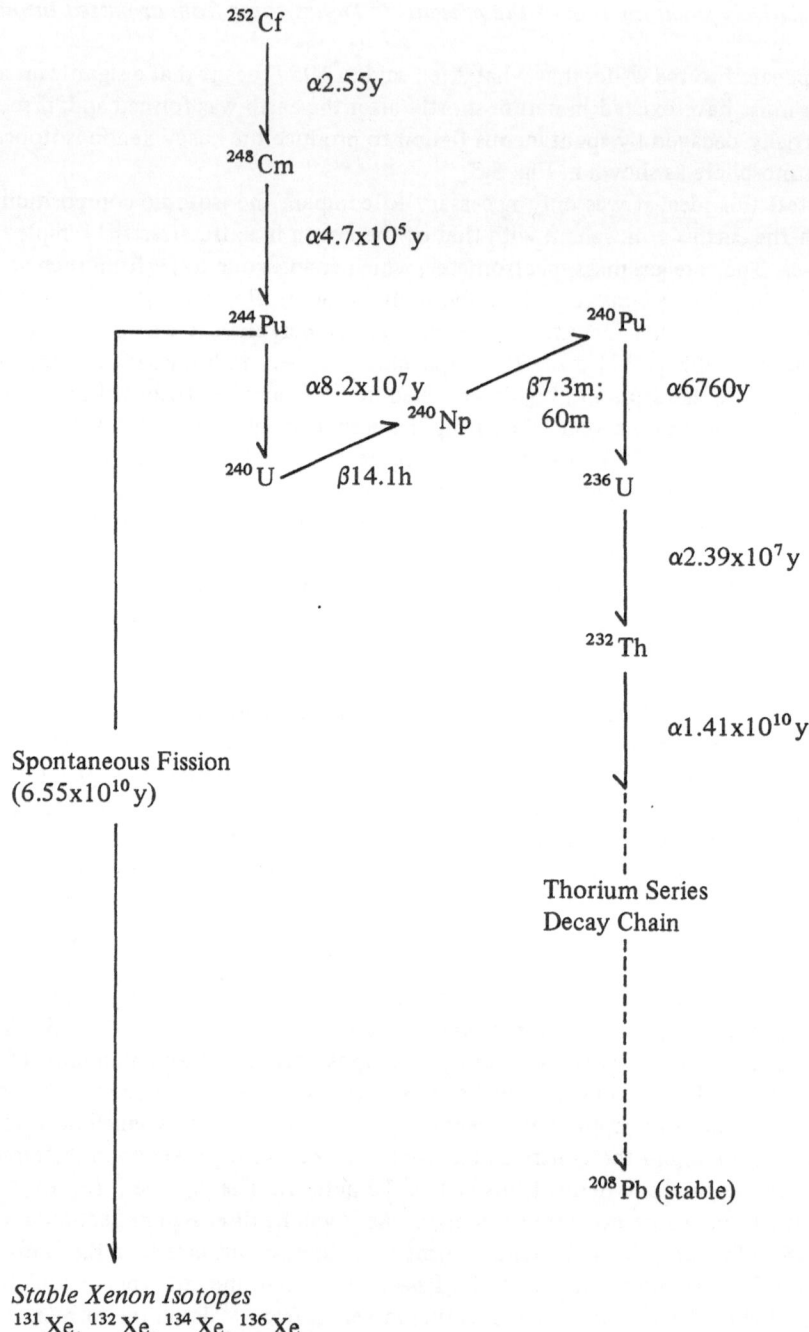

Fig. 6.3. The beginning of the thorium series decay chain

The differences in the isotopic ratios of xenon in the earth's atmosphere (T) and in the meteorite (M) may be written in the following manner:

$$\delta_i = \left(\frac{^iXe}{^{130}Xe}\right)_T - \left(\frac{^iXe}{^{130}Xe}\right)_M \tag{6.6}$$

where ^{130}Xe is used as a reference standard for the following reasons: (a) the abundance of this nuclide is fairly large, and hence the errors in the mass-spectrometric measurements are expected to be small; (b) it is shielded by ^{130}Te (the half-life for the double β^--decay of ^{130}Te was estimated by Inghram and Reynolds to be 1.4×10^{21} years), and the amount produced by fission can be regarded as negligible; (c) possibility of the production of this nuclide by other nuclear reactions appears to be comparatively small.

The values of δ_i calculated from the data by Reynolds [6.11] and Nier [6.14] are shown in Table 6.3.

Table 6.3. Comparison of the isotopic compositions of xenon in the atmosphere and in the meteorite Richardton, according to Kuroda (1960)

Mass No. (i)	Atmosphere $(^iXe/^{130}Xe)_T$	Meteorite $(^iXe/^{130}Xe)_M$	δ_i	Δ_i
124	0.024 ± 0.000	0.035 ± 0.004	−0.011 (± 0.004)	−0.0005
126	0.022 ± 0.000	0.029 ± 0.004	−0.007 (± 0.002)	−0.0003
128	0.470 ± 0.003	0.528 ± 0.082	−0.058 (± 0.085)	−0.0002
129	6.47 ± 0.06	9.18 ± 0.29	−2.71 (± 0.35)	−0.02
130	1.000 ± 0.000	1.000 ± 0.000	0.000	0.000
131	5.18 ± 0.05	5.05 ± 0.17	+0.13 (± 0.22)	+0.01
132	6.58 ± 0.05	6.15 ± 0.16	+0.43 (± 0.21)	+0.03
134	2.56 ± 0.01	2.40 ± 0.06	+0.16 (± 0.07)	+0.02
136	2.17 ± 0.01	1.98 ± 0.03	+0.19 (± 0.04)	+0.03

The large negative value of δ_i for ^{129}Xe is due to the decay:

$$^{129}I \xrightarrow{1.7 \cdot 10^7 \text{ yr}} {}^{129}Xe$$

as pointed out by Reynolds. The small negative values of δ_i for the lighter isotopes are probably due to mass fractionation or cosmic-ray-induced nuclear reactions.

Kuroda then noted that Suess [6.28] derived the following equation which correlates the ratio of the terrestrial abundance (N_{ter}) and the cosmic abundance (N_{sol}) of the rare gases with the mass (M/m_1):

$$-\log \frac{N_{ter}}{N_{sol}} = 10 \cdot e^{-0.045 M/m_1} + 7.1 \tag{6.7}$$

6.28   H.E. Suess, J. Geol. 57:600 (1949)

Differences (Δ_i) in the isotopic ratios of xenon in the earth's atmosphere and the cosmic abundance ratios $(^iXe/^{130}Xe)_C$,

$$\Delta_i = \left(\frac{^iXe}{^{130}Xe}\right)_T - \left(\frac{^iXe}{^{130}Xe}\right)_C \tag{6.8}$$

can be calculated from the Suess equation. The values of Δ_i thus are shown in column 5 of Table 6.3.

The values of δ_i and Δ_i in Table 6.3 vary in the same direction. The absolute value of δ_i is, however, always much greater than that of Δ_i, suggesting that the effect of the mass fractionation is probably small.

A maximum contribution from the ^{238}U spontaneous fission can be calculated from the terrestrial abundance ratio of uranium to xenon. The number of atoms of ^iXe produced by the ^{238}U spontaneous fission is:

$$(^iXe)_f = \frac{\lambda_f}{\lambda_\alpha + \lambda_f} \; (^{238}U)[\exp\{(\lambda_\alpha + \lambda_f)T\} - 1] \cdot Y_i \tag{6.9}$$

where λ_α and λ_f are the decay constants for the alpha decay and the spontaneous fission of ^{238}U, respectively, (^{238}U) is the number of atoms of ^{238}U in the system, Y_i is the fission yield for ^iXe, and T is the age of the system.

From equations (6.6) and (6.9), we have

$$\delta_i = \left[\left\{\frac{(^{238}U)}{(^{130}Xe)}\right\}_T - \left\{\frac{(^{238}U)}{(^{130}Xe)}\right\}_M \right] \cdot \frac{\lambda_f}{\lambda_\alpha + \lambda_f} \cdot$$
$$[\exp \; (\lambda_\alpha + \lambda_f)T - 1] \cdot Y_i \tag{6.10}$$

According to Damon and Kulp [6.29], the present rate of escape of ^4He and ^{40}Ar from the earth's crust is far too small to account for the quantity of ^{40}Ar in the atmosphere. They concluded that most of the ^{40}Ar was introduced into the atmosphere in the first billion years of earth history from the mantle and crust. The positive values of δ_i for the heavier isotopes of xenon indicate that the uranium/xenon ratio in the meteorite system is smaller than that in the terrestrial system and hence:

$$\delta_i < \left\{\frac{(^{238}U)}{(^{130}Xe)}\right\}_T \cdot \frac{\lambda_f}{\lambda_\alpha + \lambda_f} \cdot [\exp\{(\lambda_\alpha + \lambda_f)(4.5 \times 10^9)\} - \tag{6.11}$$
$$\{\exp \; (\lambda_\alpha + \lambda_f)(3.5 \times 10^9)\}] \cdot Y_i$$

Introducing the following set of values into equation (6.11), we can set an upper limit for the contribution from the ^{238}U spontaneous fission:

6.29   P.E. Damon and J.L. Kulp, Geochim. Cosmochim. Acta 13:280 (1958)

U content of the earth's crust = 1.3 ppm;
U content of the mantle = 0.01 ppm;
mass of crust = 23.7 × 10²⁴ grams;
mass of mantle = 4,060 × 10²⁴ grams; and   (6.12)
the number of atoms of xenon in the
    atmosphere = 8.14 × 10³⁶.

Equation (6.11) yields a value of $\delta_i < 0.004$ for ^{136}Xe, which is much smaller than the observed value of $\delta_i = 0.19 \pm 0.04$. This shows that the ^{238}U spontaneous fission alone

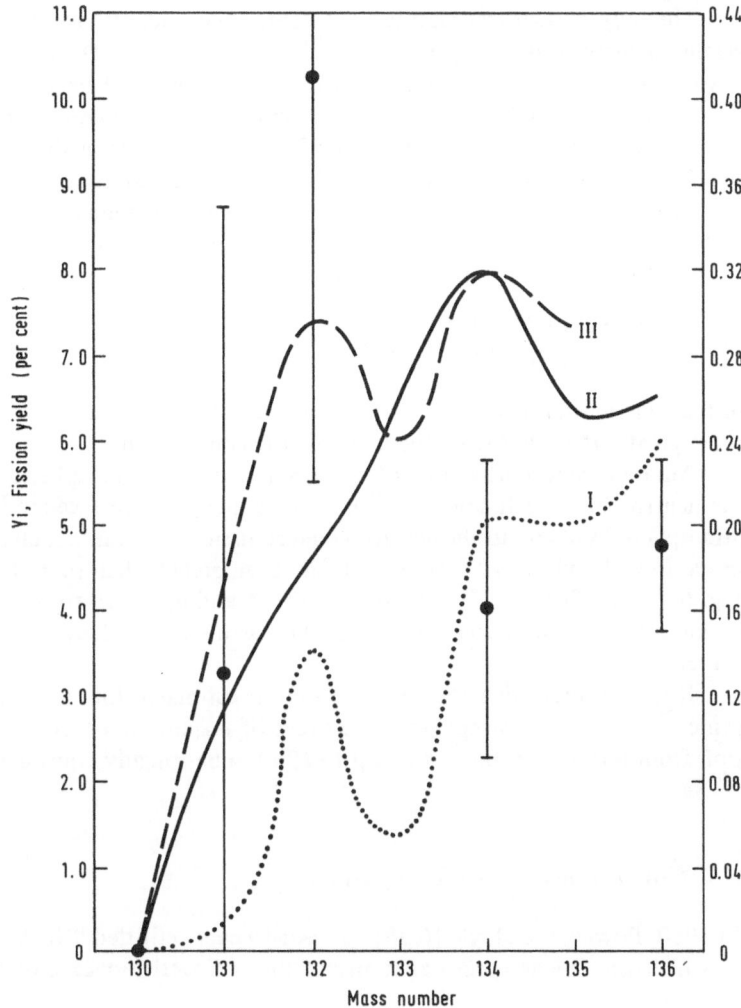

Fig. 6.4. Comparison of the value of δ_i with the mass-yield curves for various types of fission, according to Kuroda (1960). I, uranium-238 spontaneous fission; II, uranium-235 neutron-induced fission; III, curium-242 spontaneous fission. The mass-yield curve for uranium-235 neutron-induced fission was taken from the report by Katcoff [6.30], that for curium-242 spontaneous fission was taken from the data by Steinberg and Glendenin [6.31], and that for uranium-238 spontaneous fission was constructed from the xenon data by Wetherill [6.32] and from the iodine data by Kuroda, Edwards, and Ashizawa [6.33] and Ashizawa and Kuroda [6.34]

could not have caused such a large difference in the ^{136}Xe/^{130}Xe ratio between that in the atmosphere and in the meteorite.

The values of δ_i for i = 130 to 136 are plotted in Fig. 6.4 together with the mass-yield curves for the spontaneous fission of ^{238}U, ^{242}Cm, and also for the neutron-induced fission of ^{235}U.

The mass-yield curve thus obtained from the values of δ_i differs from that for the ^{238}U spontaneous fission in that the fission yield at mass 132 seems to be abnormally high. It was felt interesting that a similar peak seemed to exist in the ^{242}Cm spontaneous fission mass-yield curve. Kuroda thus concluded that the abnormally high yield at mass 132 might be due to the spontaneous fission of some of the extinct transuranium elements.

The only known transuranium element, which might have been present in an appreciable quantity in the early history of the earth is ^{244}Pu. The half-life of ^{244}Pu (8×10^7 years) is not too short as compared with the time-interval between the formation of the element and that of the earth. If it is assumed that ^{244}Pu and ^{238}U were formed in equal abundance at the time of formation of the elements, then the ^{244}Pu/^{238}U ratio at the time of the formation of the earth would have been approximately 0.05.

The ratio of the α-decay half-life and the spontaneous fission half-life of ^{244}Pu is $(8 \times 10^7)/(2.5 \times 10^{10}) = 3.2 \times 10^{-3}$, while that of ^{238}U is $(4.5 \times 10^9)/(8.0 \times 10^{15}) = 5.6 \times 10^{-7}$. Thus, the ratio of the contributions from ^{244}Pu and ^{238}U is calculated to be:

$$\frac{(0.05)\,(2)\,(3.2 \times 10^{-3})}{(5.6 \times 10^{-7})} = 570$$

or the contribution from the ^{244}Pu spontaneous fission would have been several hundred times greater than that from the ^{238}U spontaneous fission.

Another important source of contribution to the atmospheric inventory of xenon is the neutron-induced fission of ^{235}U. If the apparent excess xenon isotopes in the earth's atmosphere were due to the neutron-induced fission (or uranium chain reactions occurring in nature), the yield of ^{134}Xe should have been greater than that of ^{136}Xe (see Curve II in Fig. 6.4). The fact that $\delta_i = 0.16$ for i = 134 and $\delta_i = 0.19$ for i = 136, however, seemed to indicate that the excess fission in the earth's atmosphere was not produced by this source.

It is also interesting to note that Kuroda estimated that the total inventory of fissiogenic ^{136}Xe in the atmosphere to be order of magnitude of 10^6 ton, while the contribution from nuclear weapons tests up to 1960 to be roughly equivalent to about 0.5 ton of ^{136}Xe.

6.7. Chronology of Nucleosynthesis

In 1960, Fowler and Hoyle [6.35] published a paper entitled "Nuclear Cosmochronology", in which they made quantitative use of the radioactive decays of ^{232}Th, ^{235}U and ^{238}U

6.30   S. Katcoff, Nucleonics 16:78 (1958)
6.31   E.P. Steinberg and L.E. Glendenin, Phys. Rev. 95, 431 (1954)
6.32   G.W. Wetherill, Phys. Rev. 92:907 (1953)
6.33   P.K. Kuroda, R.R. Edwards and F.T. Ashizawa, J. Chem. Phys. 25:603 (1956)
6.34   F.T. Ashizawa and P.K. Kuroda, J. Inorg. Nucl. Chem. 5:12 (1957)
6.35   W.A. Fowler and F. Hoyle, Ann. Physics 10:280 (1960)

in cosmochronology in much the same manner as these decays have been employed in geochronology. Their calculations led to the conclusion that the age of the Galaxy is $15^{+5}_{-2} \times 10^9$ years or $11 \pm 6 \times 10^9$ years, depending on the cosmological models used in the calculations.

In 1961, Kohman [6.36] derived the following equation for the $^{129}\text{I}/^{127}\text{I}$ ratio at the time of condensation of the solar system:

$$(^{129}\text{I}/^{127}\text{I})_0 = \frac{\rho_{129}}{\rho_{127}} \cdot \frac{\lambda_*}{\lambda_{129}} \, (e^{\lambda_* \Delta} - 1)^{-1} \cdot e^{-\lambda_{129} \Xi} \tag{6.13}$$

where Ξ is the interval between the termination of the production of ^{129}I and the beginning of the retention of its daughter, ρ_{129} and ρ_{127} are the production rates of ^{129}I and ^{127}I; λ_{129} is the decay constant of ^{129}I; λ_* is the rate constant for the nucleosynthesis (the rate of nucleosynthesis is assumed to decay exponentially since the mean time of beginning with the decay constant λ_*), and Δ is the duration of nucleosynthesis.

In 1961, Kuroda [6.37] used the model of Kohman and derived the following expression for the initial $^{244}\text{Pu}/^{238}\text{U}$ at the time of cessation of nucleosynthesis:

$$\alpha = \frac{\rho_{244}}{\rho_{238}} \cdot \frac{\lambda_{238} - \lambda_*}{\lambda_{244} - \lambda_*} \cdot \frac{\exp\,(-\lambda_* \Delta) - \exp\,(-\lambda_{244}\,\Delta)}{\exp\,(-\lambda_* \Delta) - \exp\,(-\lambda_{238}\,\Delta)} \tag{6.14}$$

where ρ_{244} and ρ_{238} are the production rates of ^{244}Pu and ^{238}U, respectively, λ_* is the rate constant for the nucleosynthesis (the rate of nucleosynthesis is assumed to decay exponentially since the mean time of beginning with the decay constant λ_*), and Δ is the duration of nucleosynthesis.

If it is assumed that

$$\begin{aligned}
\rho_{244} &= \rho_{238} \\
\lambda_* &= 0.1 \times 10^{-9} \text{ year}^{-1} \\
\Delta &= 6 \times 10^9 \text{ years}
\end{aligned} \tag{6.15}$$

a value of $\alpha = 0.0235$ is obtained from equation (6.14).

The expression for the initial $^{129}\text{I}/^{127}\text{I}$ is

$$\beta = \frac{\rho_{129}}{\rho_{127}} \cdot \frac{\lambda_*}{\lambda_{129}} \cdot \frac{1}{e^{\lambda_* \Delta} - 1} \tag{6.16}$$

A value of $\beta = 3.0 \times 10^{-3}$ is obtained from equation (6.16), if it is assumed that $\rho_{129} = \rho_{127}$.

6.36   T.P. Kohman, J. Chem. Ed. 38:73 (1961)
6.37   P.K. Kuroda, Geochim. Cosmochim. Acta 24:40 (1961)

6.8. Plutonium-244 in the Early Solar System

Since the writer put forward the hypothesis in 1960 that extinct ^{244}Pu spontaneous fission may be responsible for the general Xe anomalies in the mass region 131-136, several instances have been reported of the earth's atmosphere being enriched in the heavier Xe isotopes relative to the meteorites. However, no clear-cut evidence had been obtained previously for fission produced $^{131-136}$Xe in meteorites as would be expected from the theory in high U-Th/low primordial Xe meteorites, until Marvin W. Rowe joined the writer's group as a graduate student in 1963. Rowe then decided to search for evidence of fission xenon in the high U-Th meteorites. Several meteorites with unusually high Th contents were chosen to be examined. The first sample of this group which was examined, a 4.45-gram piece of the Pasamonte eucrite, did indeed contain fissiogenic xenon [6.38], as shown in Table 6.4.

Table 6.4. Fissiogenic xenon from the Pasamonte meteorite. These data were reported by M.W. Rowe and P.K. Kuroda in Journal of Geophysical Research 70:709 (1965)

^{i}Xe/^{130}Xe	Pasamonte	Average Carbonaceous Chondrites [6.39]	Atmosphere
i = 128	0.59 ± 0.01	0.52	0.47
i = 129	6.45 ± 0.08	6.48(a)	6.48
i = 130	≡1.00	≡1.00	≡1.00
i = 131	5.22 ± 0.06	5.08	5.19
i = 132	6.85 ± 0.07	6.23	6.59
i = 134	3.26 ± 0.02	2.36	2.56
i = 136	2.92 ± 0.02	2.00	2.17

^{130}Xe = 0.35 × 10^{-11} cc STP/g

(a) Value for Murray only

The values of δ_i and Δ_i shown in Table 6.5 were calculated in the following manner:

$$\delta_i = (^iXe/^{130}Xe)_P - (^iXe/^{130}Xe)_{Atmosphere} \qquad (6.17)$$

and

$$\Delta_i = (^iXe/^{130}Xe)_P - (^iXe/^{130}Xe)_{AVCC} \qquad (6.18)$$

where the subscript P refers to Pasamonte, and AVCC refer to the xenon in the average carbonaceous chondrites. The values of Δ_i thus calculated are plotted against the mass number i in Fig. 6.5.

The relative abundance of the excess heavy isotopes found in the Pasamonte meteorite showed a striking resemblance to the general shape of the ^{238}U spontaneous fission mass-yield curve. Rowe and Kuroda [6.38] thus concluded that the excess fission xenon

---

6.38   M.W. Rowe and P.K. Kuroda, J. Geophys. Res. 70:709 (1965)

Table 6.5. Excess fission xenon in the Pasamonte meteorite

Mass number (i)	δ_i	Δ_i
128	0.12 ± 0.01	0.07 ± 0.02
129	-0.03 ± 0.08	-0.03 ± 0.09
130	$\equiv 0.00$	$\equiv 0.00$
131	0.03 ± 0.06	0.14 ± 0.08
132	0.26 ± 0.07	0.62 ± 0.12
134	0.70 ± 0.02	0.90 ± 0.10
136	0.75 ± 0.02	0.92 ± 0.10

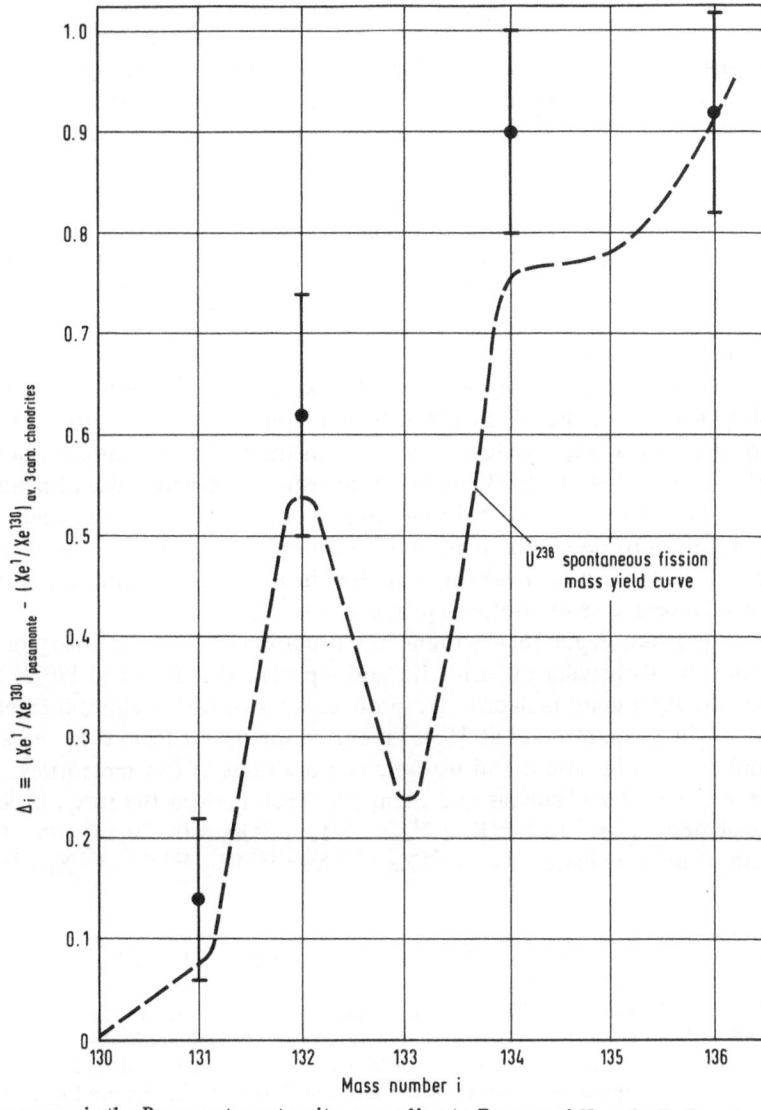

Fig. 6.5. Excess fission xenon in the Pasamonte meteorite, according to Rowe and Kuroda (J. Geophys. Res. 70:709 (1965))

was produced by the spontaneous fission of ^{238}U and ^{244}Pu in the meteorite, the contribution from the latter being much greater than that from the former.

The value of Δ_i for i = 136 can be written in the form:

$$\Delta_i = \left[\left(\frac{^{238}U}{^{130}Xe} \right)_P - \left(\frac{^{238}U}{^{130}Xe} \right)_{AVCC} \right] \cdot y_{136} \times$$

$$\left[\frac{\lambda_{f238}}{(\lambda_\alpha + \lambda_f)_{238}} \cdot (e^{\lambda_{\alpha_{238}} T} - 1) + \right.$$

$$\left. \alpha \cdot \frac{\lambda_{f244}}{(\lambda_\alpha + \lambda_f)_{244}} \cdot \left\{ e^{\lambda_{\alpha_{238}}(T + \Xi)} \cdot e^{-\lambda_{\alpha_{244}}\Xi} \right\} \right] \qquad (6.19)$$

where the symbols are analogous to those introduced previously; α is the ^{244}Pu/^{238}U ratio at the time of cessation of nucleosynthesis, which is given by equation (6.14); and T is the age of the earth (= 4.55×10^9 years) as determined by Patterson et al. [6.40].

A value of Ξ = 300 million years is obtained from equation (6.19) for the Pasamonte meteorite. Rowe and Kuroda [6.38] noted the lack of a ^{129}Xe anomaly in Pasamonte. It appeared that this was easily understandable, because the value of 300 million years corresponded to more than 17 half-lives of ^{129}I.

In 1966, Rowe and Bogard [6.41] reported the isotopic composition of xenon in the second sample of Pasamonte. Since the xenon content of this specimen was extremely low (1.1×10^{-12} cc STP/g), a much larger value of Δ_i = 3.38 ± 0.12 (at i = 136) was obtained for this sample.

Many supporting evidences for the existence of ^{244}Pu in the early solar system have since been accumulated. In 1965, Fleischer et al. [6.42] reported the existence of excess fossil fission tracks produced by ^{244}Pu in the meteorites Moore County and Toluca. In 1966, Kuroda et al. [6.43] reported the isotopic compositions of xenon and the contents of iodine and uranium in a dozen samples of achondrites. These achondrites had low concentrations of other xenon components and high concentrations of uranium. The heavy xenon isotopes were markedly enriched in these samples indicating the existence of excess fissiogenic xenon isotopes produced by ^{244}Pu.

In 1966, Pepin [6.44] found the occurrence of excess fissiogenic xenon in silicates from the Estherville mesosiderite and reported that I-Xe and Pu-Xe formation intervals for the Pasamonte meteorite are probably concordant at approximately 220 ± 40 million years. In 1966, Marti et al. [6.45] reported the results from their measurements of the Kr and Xe concentrations and isotopic compositions in five meteorites: Abee, Bruderheim, H.-Ausson, Mezö-Madaras and Stannern. They derived the isotopic composition of pure spallation Kr as ^{78}Kr : ^{80}Kr : ^{82}Kr : ^{83}Kr : ^{84}Kr = 0.179 : 0.495 : 0.765 : 1.00 : 0.63, and of pure spallation Xe as ^{124}Xe : ^{126}Xe : ^{128}Xe : ^{130}Xe : ^{131}Xe : ^{132}Xe = 0.590 : 1.00

6.39   D. Krummenacher, C.M. Merrihue, R.O. Pepin and J.H. Reynolds, Geochim. Cosmochim. Acta 26:231 (1962)
6.40   C. Patterson, G. Tilton and M. Inghram, Science 121:69 (1955)
6.41   M.W. Rowe and D.D. Bogard, J. Geophys. Res. 71:686 (1966)
6.42   R.L. Fleischer, P.B. Price and R.M. Walker, J. Geophys. Res. 70:2703 (1965)
6.43   P.K. Kuroda, M.W. Rowe, R.S. Clark and R. Ganapathy, Nature 212:241 (1966)
6.44   R.O. Pepin, J. Geophys. Res. 71:2815 (1966)

: 1.45 : 0.97 : 3.9 : 0.9. Marti et al. [6.45] then corrected the Xe fission spectrum derived by Rowe and Kuroda [6.38] from Pasamonte (see Fig. 6.11) for spallation and obtained the following composition for fission Xe.

$$^{131}Xe : {}^{132}Xe : {}^{134}Xe : {}^{136}Xe = 0.22 : 1.00 : 1.02 : 1.00$$

The isotopic composition of xenon released from the Pasamonte meteorite in stepwise heating experiments was measured by Hohenberg et al. [6.46] in 1967 and evidence for the decay of extinct ^{244}Pu and ^{129}I was found in the Kapoeta howardite by Rowe [6.47] in 1970. In 1970, Hohenberg [6.48] found exceptionally pure samples of ^{244}Pu fission xenon in the uniquely uranium-rich achondrite Angra dos Reis. In 1969, Wasserburg et al. [6.49, 6.50] observed high concentrations of ^{244}Pu fission xenon in an uranium-rich mineral whitlockite from the chondrite St. Séverin, a mineral in which an excess of fission tracks had previously been noted by Cantalaube et al. [6.51] in 1967. In 1970, Podosek [6.52] showed that ^{244}Pu fission xenon in the chondrite St. Séverin is released in correlation with a fission component implanted at uranium sites in the meteorite by neutron irradiation. He reported the abundance ratio $^{244}Pu/^{238}U$ at the time the St. Séverin meteorite began to retain xenon to be 0.0127 ± 0.0026.

The evidence that the fission xenon found in the meteorites was in fact the spontaneous fission decay products of ^{244}Pu was reported by Alexander et al. [6.53] in 1971. They used 13.0 mg of "pure" ^{244}Pu (as PuO_2) and measured the relative yields of the fissiogenic xenon isotopes mass-spectrometrically. Their results, shown in Table 6.8, agreed almost perfectly with the mass-yields deduced from the meteoritic xenon data.

In 1970, The Lunar Sample Preliminary Examination Team (LSPET) [6.54] reported the existence of excess fission xenon in lunar rock 14301. In 1972, Hutcheon and Price [6.55] reported that tracks attributed to the spontaneous fission of ^{244}Pu and ^{238}U were detected in a large whitlockite crystal in the lunar breccia 14321 from the Fra Mauro formation. Several rocks from the Fra Mauro formation (Apollo 14 mission) were dated at about 3.8 to 3.95 × 10^9 years, which made them all interesting candidates in which to search for ^{244}Pu fission tracks. Also in 1972, Crozaz et al. [6.56] reported that xenon data from stepwise heating of rock 14301 showed a large fission component that is attributed to the decay of ^{244}Pu. The total quantity of fission xenon was about 15 times that expected from ^{238}U spontaneous fission. Preliminary data on a whitlockite crystal from

6.45   K. Marti, P. Eberhardt and J. Geiss, Z. Naturforsch. 21a:398 (1966)
6.46   C.M. Hohenberg, M.N. Munk and J.H. Reynolds, J. Geophys. Res. 72:3139 (1967)
6.47   M.W. Rowe, Geochim. Cosmochim. Acta 34:1019 (1970)
6.48   C.M. Hohenberg, Geochim. Cosmochim. Acta 34:185 (1970)
6.49   G.J. Wasserburg, J.G. Huneke and D.S. Burnett, Phys. Rev. Lett. 22:1198 (1969)
6.50   G.J. Wasserburg, J.G. Huneke and D.S. Burnett, J. Geophys. Res. 74:4221 (1969)
6.51   Y. Cantalaube, M. Maurette and P. Pellas in Radioactive Dating and Methods of Low Level Counting (International Atomic Energy Agency, Vienna, 1967), p. 215
6.52   F.A. Podosek, Earth Planet. Sci. Lett. 8:183 (1970)
6.53   E.C. Alexander, Jr., R.S. Lewis, J.H. Reynolds and M.C. Michel, Science 172:837 (1971)
6.54   Lunar Sample Preliminary Examination Team (LSPET), Science 173:681 (1970)
6.55   I.D. Hutcheon and P.B. Price, Science 176:909 (1972)
6.56   G. Crozaz, R. Drozd, H. Graf, C.M. Hohenberg, M. Monnin, D. Ragan, C. Ralston, M. Seitz, J. Shirck, R.M. Walker and J. Zimmerman, Proc. Third Lunar Sci. Conf. (Suppl. 3, Geochim. Cosmochim. Acta) 2:1623 (1972)

Table 6.8. Comparison of meteoritic fission yields with ^{244}Pu fission yields

Meteorite	Fission Yield				Reference
	131	132	134	136	
(1) Pasamonte	33	93	91	≡100	K. Marti, P. Eberhardt and
	±3	±8	±2.5		J. Geiss [6.45] based on the
					data of M.W. Rowe and D.D.
					Bogard [6.41]
(2) Pasamonte	25	88.5	94	≡100	C.M. Hohenberg, M.N. Munk
	±3	±3	±5		and J.H. Reynolds [6.46]
(3) Whitlockite from	31	97	93	≡100	G.J. Wasserburg, J.G. Huneke,
St. Severin	±8	±8	±1		D.S. Burnett [6.49, 6.50]
(4) Kapoeta	26	88	91	≡100	M.W. Rowe [6.47]
	±3	±4	±5		
(5) ^{244}Pu	24.6	87.0	92.1	≡100	E.C. Alexander, Jr., R.S.
	±2.2	±3.1	±2.7		Lewis, J.H. Reynolds, M.C.
					Michel [6.53]

14301 also showed a large track excess that could be due to either ^{244}Pu or an early irradiation.

It is interesting to note here that ^{244}Pu is not completely extinct on the earth today [6.57], as it can be shown by the following simple calculation: let the ^{244}Pu/^{238}U ratio 4.6 billion years ago be x. Then the same ratio today is approximately

$$(1/2)^{56-1} \cdot x = 3 \times 10^{-17} \cdot x.$$

The total mass of the earth equals 5.977×10^{27} grams and we may assume the average uranium content of the earth to be the same as that in chondrites (10 ppb). Thus the total amount of uranium in the earth is 6×10^{19} grams. Hence the total amount of ^{244}Pu remaining on the earth today must be about

$$1,800 \cdot x \text{ (grams)}.$$

The value of x can be estimated from the observed ^{136f}Xe/^{238}U ratio in meteorites. According to Rowe and Kuroda [6.38], ^{136f}Xe in the meteorite Pasamonte is 3.2×10^{-12} cc STP/g and the uranium content is 54 ppb, and hence

$$x = 0.004$$

and the total amount of the primordial ^{244}Pu remaining on the earth today is roughly

$$(1,800)(0.004) = 7 \text{ grams}$$

6.57   P.K. Kuroda, Accounts of Chemical Research 12:73 (1979)

The total inventory of man-made ^{244}Pu in the world today is not accurately known, but it is likely to be not much greater than the amount of primordial ^{244}Pu remaining on the earth. Attempts have been made by a number of investigators to detect the primordial ^{244}Pu in terrestrial minerals, but the results [6.58, 6.59] obtained so far appear to be negative or inconclusive.

6.9. Unsolved Problems in Xenology

In 1963, Reynolds [6.60] stated: "*Xenology* means to us the detailed study of the abundance of Xe isotopes evolved from meteorites by heating or other means and the inferences that can be drawn from these studies about the early history of the meteorites and the solar system. To the classicist *xenology* means study of a strange substance, which is also appropriate". In the paper entitled Xenology, he discussed xenology in the context of theories of the origin of the heavy elements by Burbidge et al. [6.61] and by Cameron [6.62], and a theory of the origin of Xe isotope anomalies in meteorites by Kuroda [6.21] and Cameron [6.63]. He cautioned, however, that ideas in this field will require frequent revision as the experimental side of the subject developes and noted that some of these ideas have already been seriously challenged by the 1962 hypothesis of Fowler et al. [6.64] concerning the nucleosynthesis of the light elements during the early history of the solar system.

According to this hypothesis, the synthesis occurred through spallation and neutron reactions simultaneously induced in the outer layers of the planetesimals by the bombardment of high energy charged particles, mostly protons, accelerated in magnetic flares at the surface of the condensing sun. Fowler et al. [6.64] showed that the neutron flux was sufficient to produce the radioactive ^{107}Pd and ^{129}I necessary to account for the radiogenic ^{107}Ag and ^{129}Xe anomalies observed in meteorites. As it was predicted by Reynolds, the field of xenology turned out to be extremely complex and many of the problems which developed in the field of xenology during the 1960's remained unsolved through the 1970's.

In 1960, Eberhardt and Geiss [6.65] discussed the possibility that the ^{129}Xe anomaly found in meteorite xenon may have arisen from the incomplete mixing of ^{129}Xe formed by the decay of ^{129}I in the solar nebula prior to the formation and cooling of meteoritic solids. Zähringer and Gentner [6.66–6.68] also commented upon a proportionality between ^{129}Xe and primordial gases in several meteorites and suggested that this observation supported the incomplete mixing hypothesis.

6.58   P.R. Fields, A.M. Friedman, J. Milsted, J. Lerner, C.M. Stevens, D. Metta and W.K. Sabine, Nature 212:131 (1966)
6.59   D.C. Hoffman, F.O. Lawrence, J.L. Mewherter and F.M. Rourke, Nature 234:132 (1971)
6.60   J.H. Reynolds, J. Geophys. Res. 68:2939 (1963)
6.61   E.M. Burbidge, G.R. Burbidge, W.A. Fowler and F. Hoyle, Rev. Mod. Phys. 29:547 (1957)
6.62   A.G.W. Cameron, Ann. Rev. Nucl. Sci. 8:299 (1958)
6.63   A.G.W. Cameron, Icarus 1:13 (1962)
6.64   W.A. Fowler, J.L. Greenstein and F. Hoyle, Geophys. J. Roy. Astron. Soc. 6:148 (1962)
6.65   P. Eberhardt and J. Geiss, Z. Naturforsch. 15a:547 (1960)
6.66   J. Zähringer and W.Z. Gentner, Z. Naturforsch. 16a:239 (1961)

In 1961, Jeffery and Reynolds [6.69] reported that if part of the ^{127}I in meteorites is converted to ^{128}Xe by thermal neutron irradiation, the ^{128}Xe so produced showed a release pattern remarkably similar to that of anomalous ^{129r}Xe. This implies that the ^{129r}Xe resides in the same mineral sites as the ^{127}I, as it would if it were produced by the *in situ* decay of ^{129}I. Merrihue and Turner [6.70] reported in 1965 that the evidence available at that time strongly supported an *in situ* decay origin of the special ^{129}Xe anomaly in meteorites. Since the initial experiments by Jeffery and Reynolds [6.69], work has been carried out on irradiated samples of Richardton [6.60], Renazzo [6.71], Bruderheim and a Bruderheim chondrule [6.72], Pantar (light and dark) and Bjurböle [6.73].

Merrihue and Turner [6.70] pointed out that in all cases there was a good correlation if only the high temperature xenon fractions were considered, a fact which implied an ^{127}I to ^{129r}Xe correlation, as expected if ^{127}I and ^{129}I were incorporated together into meteorites. If iodine and tellurium were concentrated in the same minerals, as suggested by Goles and Anders [6.74], the correlations were also consistent with a ^{129}Xe-tellurium correlation, as might be expected if ^{129}I derived partly from neutron capture in ^{128}Te, as proposed by Fowler et al. [6.62].

Merrihue and Turner [6.70] also pointed out that the *in situ* decay hypothesis was strengthened even more when one considered all of the meteorites analyzed. The meteorites Renazzo, Bruderheim, Bruderheim chondrule, Bjurböle, Pantar dark and Pantar light were all irradiated together (but in separate sealed quartz ampules), and thus received essentially the same flux of neutrons. It was found that all six meteorites showed roughly similar correlations, and the ratios of ^{129r}Xe to ^{127}I calculated from the high temperature ($\geqslant 900\ °C$) xenon fractions for five of the six samples were the same within ± 17 percent. Such a spread would correspond on the *in situ* decay hypothesis to a range of formation invervals for these meteorites of 7 million years. Merrihue and Turner [6.70] thus noted that they could not find any explanation for this relative constancy of the anomalous $^{129r}Xe/^{127}I$ ratio on the incomplete mixing hypothesis.

While it appeared that the experimental evidences favored the *in situ* decay hypothesis for the origin of anomalous ^{129}Xe in meteorites, the question of whether the extinct nuclide ^{129}I was synthesized in the *r*-process nucleosynthesis occurring in a supernova explosion or wether it was produced by neutron-capture reactions in ^{128}Te during an early irradiation of the solar nebula as proposed by Fowler et al. [6.62] remained unanswered for many years.

In 1961, Goles and Anders [6.75] suggested that the time interval (Ξ) between the cessation of nucleosynthesis and formation of the meteorites could be calculated by taking

6.67   J. Zähringer, Z. Naturforsch. 17a:460 (1962)
6.68   J. Zähringer, Ann. Rev. Astron. Astrophys. 2:121 (1964)
6.69   P.M. Jeffery and J.H. Reynolds, J. Geophys. Res. 66:3582 (1961); see also Z. Naturforsch. 16a:431 (1961)
6.70   C. Merrihue and G. Turner, Z. Naturforsch. 20:961 (1965)
6.71   J.H. Reynolds and G. Turner, J. Geophys. Res. 69:3263 (1964)
6.72   C. Merrihue, J. Geophys. Res. 71:263 (1966)
6.73   G. Turner, unpublished manuscript (1965)
6.74   G.G. Goles and E. Anders, Geochim. Cosmochim. Acta 26:723 (1962)

advantage of the fact that two extinct nuclides, ^{244}Pu and ^{129}I, decay to isotopes of the same element. The results from all previous investigations were either inconclusive or contradictory, chiefly because the abundances of iodine and uranium and the isotopic compositions of xenon had not previously been measured, as part of the same investigation, in a set of samples from a reasonably large number of meteorites.

In a paper entitled "Galactic and Solar Nucleosynthesis" published in 1966, Kuroda et al. [6.43] reported that almost all the meteorites so far investigated seemed to fit an "isochron" corresponding to the ^{244}Pu-^{136}Xe formation interval $\Xi = 300 \times 10^6$ years, whereas the values of the ^{129}I-^{129}Xe formation interval tended to scatter more or less randomly within the range $100-300 \times 10^6$ years. They concluded that these rather striking results could be interpreted on the basis of the postulate that ^{244}Pu is synthesized solely in the supernova explosion, while ^{129}I can be synthesized in the solar, as well as in the galactic nucleosynthesis processes.

In 1967, Sabu and Kuroda [6.76] reported, however, that they examined all the xenon isotope ratio data obtained in their laboratory and elsewhere prior to 1967, and re-calculated the Pu/Xe and I/Xe decay intervals of many meteorites. Concordant decay intervals were obtained for a dozen meteorites, mostly achondrites, and they concluded that ^{244}Pu and ^{129}I abundances in the early solar system suggested that these extinct nuclides were synthesized in the galactic nucleosynthesis process, which lasted several billion years. In 1968, Sabu and Kuroda [6.77] reported that the Bruderheim chondrite contained about 1×10^{-12} cc STP ^{136f}Xe, and none of the ordinary chondrites studied so far seemed to contain much greater amounts of fissiogenic xenon than this. From the experimental data obtained in different laboratories the following average values were secured for the Bruderheim chondrite: ^{136f}Xe $= 1.06 \times 10^{-12}$ cc STP/g; ^{129r}Xe $= 41 \times 10^{-12}$ cc STP/g; I = 19 ppb; and U = 11 ppb.

The above values yield the Pu/Xe and I/Xe decay intervals of 142×10^6 and 136×10^6 years, respectively, if the initial ^{244}Pu/^{238}U and ^{129}I/^{127}I ratios immediately after cessation of nucleosynthesis were 0.0235 and 0.0030, respectively, which can be obtained from equations (6.14)–(6.16).

In 1968, Reynolds [6.78] reported, however, that he was unable to find concordant decay intervals of the sort reported by Sabu and Kuroda [6.74] even when working with the same data. He felt that a detailed concordancy of the Pu/Xe and I/Xe formation intervals was unlikely to be demonstrated with existing data for a number of reasons: (a) there is the known problem of ^{129}Xe loss, and fissiogenic ^{136}Xe is probably also lost from some sites; (b) there are known variations, sometimes by as large as a factor of two, in the concentrations of uranium and iodine in different samples of the same meteorites; and (c) the assumption that U and Pu have not been fractionated is open to serious question. Reynolds [6.78] added, however, that he did not conclude from his work that there is no concordancy between the I/Xe and Pu/Xe methods of dating achondrites. He felt that the methods as applied thus far to achondrites were still too rough to merit any such conclusion (see also the report by Kuroda [6.79]).

6.75   G.G. Goles and E. Anders, J. Geophys. 66:889 (1961)
6.76   D.D. Sabu and P.K. Kuroda, Nature 216:442 (1967)
6.77   D.D. Sabu and P.K. Kuroda, J. Geophys. Res. 73:3957 (1968)
6.78   J.H. Reynolds, Nature 218:1024 (1968)
6.79   P.K. Kuroda, Nature 221:726 (1969)

The opinion expressed by Sabu and Kuroda [6.76, 6.77] did not seem to be in line with the conclusion reached by Hohenberg et al. [6.80] concerning the so-called sharp-isochronism observed in chondrites. They reported in 1967 that measurements of the accumulation of ^{129}Xe from radioactive decay of extinct ^{129}I in meteorites showed that the ^{129}I/^{127}I ratio in high-temperature minerals in diverse chondrites was 10^{-4} at the time of cooling. According to these investigators, the uniformity in the ratio indicated that the minerals cooled simultaneously within 1 or 2 million years. Hohenberg and Reynolds [6.81] reported in 1969 that iodine-xenon dating depends for its validity not upon quantitative retention but upon what is effectively proportionate iodine-xenon retention. They reasoned that one model which exhibits this behavior would have the retentively sited iodine and radiogenic ^{129}Xe co-existing at trapping sites, which undergo thermal rupture as heating proceeds. Until thermal rupture, the trapping sites retain both iodine and xenon quantitatively.

In 1970, Podosek [6.82] found, however, that the ^{129}I/^{127}I ratio for several chondrites determined by the method of Hohenberg et al. [6.80] was not quite as constant as it was expected to be. He expressed the results in terms of relative ages arbitrarily referred to the chondrite Bjurböle, which had the ^{129}I/^{127}I ratio of 1.09×10^{-4} (atom/atom). Wetherill [6.83], in 1975, prepared an updated version of the figure of Podosek [6.82]. On this scale the primitive Orgueil, Murchison, and Allende material have ages of -5.7, -5.5, and -2.4 million years, which means that they were formed at least a few million years prior to the Bjurböle chondrite. Metamorphosed chondrites such as Karoonda also appear, however, as early as -3.9 million years.

In 1970, Podosek and Hohenberg [6.84] also reported that precise data on the unequilibrated chondrite Manych, which contained large excesses of ^{129}Xe, clearly demonstrated a lack of correlation between the radiogenic ^{129}Xe and iodine. In 1976, Drozd and Podosek [6.85] published a paper entitled "Primordial ^{129}Xe in Meteorites", in which they reported that the I/Xe formation age of the Arapahoe chondrite was the oldest yet observed, 9.9 ± 0.8 million years before that of Bjurböle. Moreover, they reported that the composition of trapped xenon in Arapahoe was normal except for a deficiency of ^{129}Xe, where they inferred ^{129}Xe/^{132}Xe = 0.56 ± 0.04, well below the apparent primordial solar system value. (For a more detailed discussion on the subject of I-Xe dating, see the papers by Kuroda [6.86] and Podosek [6.87, 6.88].)

6.80   C.M. Hohenberg, F.A. Podosek and J.H. Reynolds, Science 156:202 (1967)
6.81   C.M. Hohenberg and J.H. Reynolds, J. Geophys. Res. 74:6679 (1969)
6.82   F.A. Podosek, Geochim. Cosmochim. Acta 34:341 (1970)
6.83   G.A. Wetherill, Ann. Rev. Nucl. Sci. 25:283 (1975)
6.84   F.A. Podosek and C.M. Hohenberg, Earth Planet. Sci. Lett. 8:443 (1970)
6.85   R.J. Drozd and F.A. Podosek, Earth Planet. Sci. Lett. 31:15 (1976)
6.86   P.K. Kuroda, Geochim. J. 10:65 (1976)
6.87   R.J. Drozd and F.A. Podosek, Geochem. J. 11:231 (1977)
6.88   T.J. Bernatowicz and F.A. Podosek, in: Terrestrial Rare Gases, edited by E.C. Alexander, Jr. and M. Ozima, Japan Scientific Societies Press, 1978, p. 99

Several investigators have determined the $^{244}Pu/^{238}U$ ratio at the time meteorites began retaining xenon. For example, Podosek [6.89], in 1970, reported that the $^{244}Pu/^{238}U$ ratio for the St. Séverin meteorite was 0.0127 ± 0.0026. He also reported that the $^{129}I/^{127}I$ ratio for the St. Séverin meteorite was $(0.81 \pm 0.03) \times 10^{-4}$. It is interesting to note that these ratios yield concordant Pu/Xe and I/Xe decay intervals, if the values of α and β calculated from equations (6.14) and (6.16) are used. A value of $(83 \pm 1) \times 10^6$ years is obtained for the I/Xe decay intervals, while a value of $(74 \pm 24) \times 10^6$ years is obtained for the Pu/Xe decay intervals.

Podosek and Lewis [6.90], in 1972, reported, however, that the $^{244}Pu/^{238}U$ ratio at the time the Allende meteorite started to retain xenon was 0.087 ± 0.011. This ratio is nearly six times higher than that found in the chondrite St. Séverin and three times the value of $\alpha = 0.0235$ calculated from equation (6.14), assuming the conditions given by equation (6.15) to hold. According to Podosek and Lewis [6.90], this result must be ascribed to chemical fractionation between Pu and I in the formation of these inclusions.

Aside from the problems encountered in I-Xe and Pu-Xe dating methods, a puzzle concerning the origin of the so-called carbonaceous chondrite fission (CCF) xenon remained unsolved for more than a decade. In 1964, Reynolds and Turner [6.91] reported the presence of a fission xenon in the carbonaceous chondrite Renazzo. This xenon component later became widely known as the Renazzo-type fission xenon or CCF. Many papers have been published on the subject of CCF, xenon and the problems related to CCF will be discussed in Chapter 7, Isotopic Anomalies in the Early Solar System. Several papers have been published also on the subject of Pu/Xe and I/Xe decay intervals in relation to the chronology of nucleosynthesis, but a discussion on this interesting topic is beyond the scope of this monograph. The readers are referred to the original papers by Fowler [6.92, 6.93], Wasserburg et al. [6.94, 6.95] and Hohenberg [6.96].

It is important to note here that an interesting new idea was presented by Clayton [6.97, 6.98] in 1975. He pointed out that the xenon anomalies trapped in meteorites and the moon may have first been trapped in circumstellar grains formed in or outside of post-explosive stars. In that case, the initial solar nebula need not have contained most of their radioactive progenitors, and this would necessitate major revision of the history of the solar system formation.

As mentioned earlier, Jeffrey and Reynolds [6.69] performed stepwise release of anomalous ^{129}Xe and showed that it was correlated with the ^{128}Xe created by neutron irradiation of ^{127}I within the meteorite samples. This demonstration, showing that the anomalous ^{129}Xe was linearly associated with iodine in the meteorites, refuted the sug-

6.89   F.A. Podosek, Earth Planet. Sci. Lett. 8:183 (1970)
6.90   F.A. Podosek and R.S. Lewis, Earth Planet. Sci. Lett. 15:101 (1972)
6.91   J.H. Reynolds and G. Turner, J. Geophys. Res. 69:3263 (1964)
6.92   W.A. Fowler, Cosmology, Fusion and Other Matters (George Gamow Memorial Volume) edited by Frederick Reines, Colorado Associated University Press, 1972, p. 67
6.93   W.A. Fowler, "Proceedings of The Robert A. Welch Foundation Conference on Chemical Research XXI. COSMOCHEMISTRY", November 7–9, 1977, Houston, Texas, p. 61
6.94   G.J. Wasserburg, D.N. Schramm and J.C. Huneke, Astrophys. J. 157:L91 (1969)
6.95   G.J. Wasserburg, J.C. Huneke and D.S. Burnet, Phys. Rev. Letters 22:1198 (1969)
6.96   C.M. Hohenberg, Science 166:212 (1969)
6.97   D.D. Clayton, Astrophys. J. 199:765 (1975)
6.98   D.D. Clayton, Nature 257:36 (1975)

gestion of Eberhardt and Geiss [6.65] that the decay was not *in situ* but was instead accreted ^{129}Xe-rich gas from a xenon-rich atmosphere. This argument had since been taken as proof that ^{129}I existed in the solar system and that the ^{129}Xe anomaly measured the relative solidification times of the meteorites. According to Clayton [6.97], however, this last conclusion had not been totally justified, because the stepwise-release experiments are consistent with ^{129}I decay to ^{129}Xe in interstellar grains long before the solar system existed.

In 1976, Begemann and Stegmann [6.99] reported, however, that according to the hypothesis of Clayton, anomalies should be observed in the isotopic composition of numerous elements. For example, ^{41}K is synthesized predominantly as ^{41}Ca (half-life, 1.3×10^5 years). Therefore, in an early condensate from the solar nebula, enriched in Ca relative to K, ^{41}K should be overabundant provided that no homogenization of potassium occurred later on. Begemann and Stegmann [6.96] found, however, the isotopic composition of potassium from the Allende inclusion to be indistinguishable from that of terrestrial potassium.

6.10. Search for Superheavy Elements in Nature

There has been considerable interest among nuclear physicists and chemists in the possibility of discovering superheavy elements during the past decade [6.100, 6.101]. Theoretical predictions of the properties of superheavy nuclei have been made by several groups of investigators [6.102–6.105]. The results from these studies indicate that some superheavy elements might have a half-life as long as 10^9 years. The nucleus

$$^{298}_{114}[\quad]_{184}$$

is expected to be particularly stable because of the closing of both a proton shell and a neutron shell occurs at this location. The prediction by Nilsson et al. [6.102] that the half-life of the nucleus

$$^{294}_{110}[\quad]_{184}$$

6.99   F. Begemann and W. Stegmann, Nature 259:549 (1975)
6.100   Glenn T. Seaborg, Annual Review of Nuclear Science 18:53 (1968)
6.101   S.G. Thompson and C.F. Tsang, Science 178:1047 (1972)
6.102   S.G. Nilsson, C.F. Tsang, A. Sobiczewski, Z. Czymański, S. Wycech, C. Gustafson, I.L. Lamm, P. Möller and B. Nilsson, Nucl. Phys. A 131:1 (1969)
6.103   M. Bolsterli, E.O. Fiset, J.R. Nix and J.L. Norton, Phys. Rev. Lett. 27:681 (1971); Phys. Rev. C 5:1050 (1972)
6.104   J. Grumann, U. Mosel, B. Fink and W. Greiner, Z. Phys. 228:371 (1969)
6.105   M. Brack, J. Damgaad, A. Stenholm-Jensen, H.C. Pauli, V.M. Strutinsky and C.Y. Wong, Rev. Mod. Phys. 44:320 (1972)

(eka-platinum) should be in the neighborhood of 10^8 years, suggested that small amounts of superheavy elements might still be present in nature. Attempts have been made by several groups of investigators to detect the superheavy elements in terrestrial minerals, but the results obtained so far appear to be negative or inconclusive. As pointed out by Thompson and Tsang [6.101], however, these theoretical calculations involve great uncertainties. Thus the prediction of a half-life of 10^9 years may be uncertain by a factor of 10^6 either way; that is, the half-life may well be anything between 10^3 and 10^{15} years.

It is quite possible that the half-lives of the superheavy elements are actually much shorter than the values predicted by the above investigators. In 1976, Reichstein and Malik [6.106] calculated the potential energy surface for spontaneous fission using realistic density distributions for finite nuclei, and reported their predicted upper limit for the spontaneous fission half-lives of element 112 and 114 to be only one year.

Although a number of reports on the natural occurrence of superheavy elements have appeared in recent years, none of the experimental evidence presented so far turned out to be conclusive [6.107–6.113]. In 1973, Fleischer and Hart [6.114] reported an alternative explanation of the fossil tracks in lunar minerals that have been attributed by Bhandari et al. [6.111, 6.112] to extinct radioactivity of superheavy elements. Fleischer and Hart [6.114] showed that the longer tracks observed by Bhandari et al. [6.111, 6.112] and attributed to fission of ^{244}Pu and superheavy nuclei can be explained as normal, unaltered cosmic ray Fe or Ni tracks that have been formed since the last shock event that produced shortening of tracks. Also in 1973, Geisler et al. [6.115] reported that they used an improved method of measuring low levels of spontaneous fission to establish limits for postulated superheavy elements in natural materials. The apparatus was also used to give an approximate measure of the rate of cosmic ray-induced fission of lead, and their result for lead agreed with theoretical predictions and satisfactorily explained observations of fission events in Pb glass that had previously been attributed to superheavy element decay by Flerov et al. [6.107]. Geisler et al. [6.115] also reported that they were unable to confirm the results of the experiments by Marinov et al. [6.110].

It is interesting to note here that Hoffman et al. [6.59], in 1971, reported the observation of ^{244}Pu existing in nature when they isolated 2×10^7 atoms of ^{244}Pu from bastnaesite-rich ore from Mountain Pass, California. Fleischer and Naeser [6.116] found, how-

6.106 I. Reichstein and F.B. Malik, Annals of Physics 98:322 (1976)
6.107 G.N. Flerov and V.P. Perelygin, Atomnaya Energiya 26:520 (1969) [Sov. At. Energy 26:603 (1969)]
6.108 A.G. Popeko, N.K. Skobelev, G.M. Ter-Akopyan and G.N. Gontcharov, Preprint of the Joint Institute for Nuclear Research: Dubna, 1974
6.109 G.N. Flerov et al., Preprint of the Joint Institute for Nuclear Research, Dubna, 1977
6.110 A. Marinov, G.J. Batty, A.I. Kilvington, J.L. Weil, A.M. Friedman, G.W.A. Newton, V.J. Robinson, D.J. Hemingway and D.S. Matner, Nature 234:212 (1971)
6.111 N. Bhandari, S.G. Bhat, D. Lal, G. Rajagopalan, A.S. Tamhane and V.S. Venkatavaradan, Nature 230:219 (1971)
6.112 N. Bhandari, S.G. Bhat, D. Lal, G. Rajagopalan, A.S. Tamhane and V.S. Venkatavaradan, Second Lunar Sci. Conf., Geochim. Cosmochim. Acta Suppl. 2, 3:2599 (1971)
6.113 R.V. Gentry, T.A. Cahill, N.R. Fletcher, H.C. Kaufmann, L.R. Medsker, J.W. Nelson and R.G. Flocchini, Phys. Rev. Lett. 37:11 (1976)
6.114 R.L. Fleischer and H.R. Hart, Jr., Nature 242:104 (1973)
6.115 F.H. Geisler, P.R. Phillips and R.M. Walker, Nature 244:428 (1973)
6.116 R.L. Fleischer and C.W. Naeser, Nature 240:465 (1972)

ever, that bastnaesite, a rare earth fluorocarbonate, from the Precambrian Mountain Pass deposit had an apparent Cretaceous fission track age, and hence did not reveal any anomalous fission tracks due to ^{244}Pu.

6.11. Superheavy Elementary Particles and Quarks in Nature

In a paper entitled "Chemical Signatures for Superheavy Elementary Particles", which appeared in the 7 August 1981 issue of *Science*, Cahn and Glashow [6.117] reported that models of unified fundamental interactions suggested the existence of many particles in the mass range 10×10^9 to 100×10^{12} electron volts and some of these superheavy charged particles $X^{\pm}$ may be stable or nearly so. They noted that chemical isolation of naturally occurring technetium, promethium, actinium, protactinium, neptunium, or americium would indicate the presence of superheavy particles in the forms RuX^-, SmX^-, $^{232}ThX^-$, $^{235,\,236,\,238}UX^-$, $^{244}PuX^-$, or $^{247}CmX^-$.

In a paper entitled "Negative exotic particles as low-temperature fusion catalysts and geochemical distribution", Jørgensen [6.118] suggested that minute concentrations of unfamiliar particles (X^+ or X^-) with atomic weights M between 10^2 and 10^6 (10^{11} and 10^{15} eV) may be remnants of the Big Bang and it may be possible to concentrate the isotopes containing these unfamiliar particles. He also pointed out the possibility that the neutral adducts pX and ^{2}DX, which may form tiny molecules with a second proton, or a polymer with a density $M \times 10^{10}$ g cm^{-3}, could act as a low-temperature fusion catalyst and explain the excess heat irradiated by Jupiter.

As has been described in Chapter 3, Elements 43 and 61 in Nature, Sections 3.9. and 3.11., ^{99}Tc and ^{147}Pm occur in pitchblende at concentrations of 10^{-12} and 10^{-17}, respectively [6.119, 6.120]. It would be interesting to know whether the elements 43 and 61 in the uranium ores and minerals are accompanied by RuX^- and SmX^-.

In 1964, Gell-Mann [6.121] proposed that baryons consist of three quarks:

u-quark: charge $+2/3$, strangeness 0
d-quark: charge $-1/3$, strangeness 0
s-quark: charge $-1/3$, strangeness -1.

According to Feynman [6.122], the proton should have the structure (duu) and the neutron (ddu). It is therefore possible that protons and neutrons are no longer to be considered as elementary particles. Jørgensen [6.123, 6.124] have recently discussed the problems of predicting the chemical properties of systems containing unsaturated quarks, which may be considered as being elements with $Z = \pm 1/3$.

6.117 R.N. Cahn and S.L. Glashow, Science 213:607 (1981)
6.118 C.K. Jørgensen, Nature 292:41 (1981)
6.119 B.T. Kenna dn. P.K. Kuroda, J. Inorg. Nucl. Chem. 23:142 (1961); 26:493 (1964)
6.120 M. Attrep, Jr., and P.K. Kuroda, J. Inorg. Nucl. Chem. 30:699 (1968)
6.121 M. Gell-Mann, Phys. Rev. Letters 8:214 (1964)
6.122 R.P. Feynman, Science 183:601 (1974)
6.123 C.K. Jørgensen, Structure and Bonding 34:19 (1978)
6.124 C.K. Jørgensen, Structure and Bonding 43:1 (1981)

7.
Isotopic Anomalies in the Early Solar System

"If our inconceivable ancient Universe even had any be-
ginning, the conditions determining that beginning must
now be engraved in the atomic weights."

Theodore W. Richards, December 6, 1919

The origin of the isotopic anomalies observed in meteorites appears to be attributable to the x-process nucleosynthesis, which was responsible for the production of the deficient elements lithium, beryllium and boron during the earliest stage of the history of the solar system.

7.1. The Origin of Lithium, Beryllium and Boron

In 1923, Goldschmidt [7.1] pointed out that the light elements lithium, beryllium and boron are "deficient" in the solar system and he suggested that this deficiency must have a common cause for all three elements (mass number 6, 7, 9, 10 and 11) which might be sought in some instability in certain nuclear processes (see Chapter 5, Section 5.3). Six years later, Atkinson and Houtermans [7.2] reported that the five isotopes are destroyed as a result of collisions with thermally accelerated protons in stellar atmospheres. The questions as to exactly how and where these elements were synthesized in the universe remained unanswered, however, for many years.

In 1955, Hayakawa [7.3] put forward the hypothesis that the production of cosmic rays at stellar surfaces may give rise to an appreciable nuclear transformation so as to alter

---

7. 1   V.M. Goldschmidt, Geochemistry, edited by Alex Muir, Oxford at the Clarendon Press, 1954, p. 73

7. 2   R. d'E. Atkinson and F.G. Houtermans, Z. f. Physik 54:656 (1929)

7. 3   S. Hayakawa, Prog. Theor. Phys. 13:464 (1955)

the chemical composition of stars. He pointed out that lithium, beryllium and boron are expected to be produced through the bombardment of nuclei heavier than boron by the cosmic rays of GeV energies with the formation cross sections of the order of 10 milli-barns. He also noted that the cosmic rays are produced by the sun and more efficiently by supernovae.

In 1955, Fowler et al. [7.4] stated that the synthesis of lithium, beryllium, boron and deuterium could not be accounted for in terms of any processes occurring in the interiors of normal stars and Burbidge et al. [7.5], in 1957, attributed the formation of these elements to an "x process" (see Chapter 5, Section 5.12.). Then, five years later in 1962, Fowler et al. [7.6] proposed a new theory for the production of the light elements according to which the synthesis occurred during an intermediate stage in the early history of the solar system.

According to this theory, the planetary material became largely separated, but not completely, from the hydrogen which was the main constituent of primitive solar material. Solid planetesimals of dimensions from 1 to 50 meters consisting of silicates and oxides of the metal embedded in an *icy* matrix were formed and the synthesis occurred through spallation and neutron reactions simultaneously induced in the outer layers of the planetesimals by the bombardment of high energy charged particles, mostly protons, accelerated in magnetic flares at the surface of the condensing sun.

Fowler et al. [7.6] estimated that 10 percent of terrestrial-meteoritic material was irradiated with a thermal neutron flux of $10^7/cm^2$ for an interval of 10^7 years. The neutron flux was such that it was sufficient to produce the radioactive ^{107}Pd and ^{129}I necessary to account for the radiogenic ^{107}Ag and ^{129}Xe anomalies observed in meteorites.

If an early irradiation of the planetesimals had occurred as described above, it was felt that an alteration of the isotopic composition of gadolinium in meteorites would be the one that is most easily detectable, because the thermal neutron-capture cross-sections of some of the gadolinium isotopes were known to be extremely large: the values for ^{155}Gd and ^{157}Gd are 6.1×10^4 and 2.55×10^5 barns, respectively.

In 1963, Murthy and Schmitt [7.7] reported, however, that meteoritic and terrestrial gadolinium had the same isotopic composition to within about one percent. Seven years later in 1970, Eugster et al. [7.8] determined the isotopic composition of gadolinium in a number of meteorites and reported that all samples except the Norton County achondrite showed the same relative isotopic abundances as terrestrial gadolinium. These results seemed to set an upper limit of 3×10^{15} neutrons per cm^2 on a differential integrated thermal neutron irradiation of the earth and these meteorites. Neutron-capture effects were found in gadolinium extracted from the Norton County achondrite, but Eugster et al. [7.8] concluded that these most probably have been produced by secondary neutrons during a long cosmic-ray exposure of this large stone meteorite. The observed isotopic anomalies corresponded to an integrated thermal neutron flux of $(6.3 \pm 0.9) \times 10^{15}$ neutrons per cm^2. The results from these studies on the isotopic composition of Gd in meteorites thus seemed to leave little room for an intense irradiation of solid bodies in the early solar system.

7. 4   W.A. Fowler, G.R. Burbidge, and E.M. Burbidge, Astrophys. J. Suppl. 2:167 (1955)
7. 5   E.M. Burbidge, G.R. Burbidge, W.A. Fowler and F. Hoyle, Rev. Mod. Phys. 29:547 (1957)
7. 6   W.A. Fowler, J.L. Greenstein and F. Hoyle, Geophys. J. Royal Astron. Soc. 6:148 (1962)
7. 7   V.R. Murthy and R.A. Schmitt, J. Geophys. Res. 68:911 (1963)
7. 8   O. Eugster, F. Tera, D.S. Burnett and G.J. Wasserburg, J. Geophys. Res. 75:2753 (1970)

The exact nature of the x-process is not yet fully understood. A variety of mechanisms have since been suggested for the production of the deficient elements, primarily involving the breakup of carbon, nitrogen, and oxygen through nuclear reactions with protons and alpha particles. Proposed processes include (a) reactions in regions near the stellar surface [7.9, 7.10], (b) reactions between galactic cosmic rays and nuclei in interstellar matter [7.11–7.14] and (c) reactions in supernova shock waves [7.15, 7.16]. In 1974, Jacobs et al. [7.17] reported that analyses of comparative production cross sections from CNO targets indicated that it was possible to match the solar system abundances of lithium, beryllium and boron with reactions dominated by protons of energies below about 25 MeV.

In a paper published in 1973, Reeves et al. [7.14] concluded that the galactic cosmic rays could spallogenically produce ^{6}Li, ^{9}Be, and ^{10}B as well as the bulk of ^{11}B and about 10 percent of the Li. It appears, however, that the deuterium can only be produced pregalactically either in the Big Bang or in some pregalactic event, which will also produce ^{3}He and ^{4}He and some ^{7}Li. According to this model, the last burst of nucleosynthesis recorded in the meteorites occurred in the galaxy, not in the solar nebula.

7.2. Isotopic Anomalies in Meteorites

In 1973, Clayton et al. [7.18] found that the high temperature phases in carbonaceous chondrites contained oxygen with anomalous isotopic composition: the proportion of the most abundant isotope ^{16}O was enriched by up to 5 percent, while the ratio of ^{17}O/^{18}O remained essentially constant in these mineral phases. Nuclear reactions within the solar nebula or solar system appeared to have been inadequate to cause so large an isotopic perturbation in such an abundant element and hence these investigators have attributed the ^{16}O excesses to galactic nucleosynthesis processes which occured in stars. It is worthy of note here that the discovery of the oxygen anomalies occurred in the same year 1973, when the above-mentioned paper by Reeves et al. [7.14] on the origin of lithium, beryllium and boron was published.

Variations in the isotopic composition of magnesium in the Allende meteorite which were not attributable to isotopic fractionation were discovered by Lee and Papanastassiou [7.19] and Gray and Compston [7.20], in 1974. In 1977, Lee et al. [7.21] found large excesses of ^{26}Mg of up to 10 percent in different phases of a Ca-Al rich inclusion in the Allende meteorite. They concluded that the data provided definitive evidence for the pre-

7. 9   R. Bernas, E. Gradsztajn, H. Reeves, and E. Schatzman, Ann. Phys. (N.Y.) 44:426 (1967)
7.10   S. Hayakawa, Prog. Theor. Phys., Extra Number: 156 (1968)
7.11   H. Reeves, W.A. Fowler, and F. Hoyle, Nature 226:727 (1970)
7.12   M. Meneguzzi, J. Audouze and H. Reeves, Astron. and Astrophys. 15, 337 (1971)
7.13   H.E. Mitler, Astrophys. and Space Sci. 17:186 (1972)
7.14   H. Reeves, J. Audouze, W.A. Fowler, and D.N. Schramm, Astrophys. J. 179:909 (1973)
7.15   A.G.W. Cameron, S.A. Colgate, and L. Grossman, Nature 243:204 (1973)
7.16   J. Audouze and J.W. Truran, Astrophys. J. 182:839 (1973)
7.17   W.W. Jacobs, D. Bodansky, D. Chamberlin and D.L. Oberg, Phys. Rev. C 9:2134 (1974)
7.18   R.N. Clayton, L. Grossman and T.K. Mayeda, Science 182:485 (1973)
7.19   T. Lee and D.A. Papanastassiou, Geophys. Res. Letters 1:225 (1974)
7.20   C.M. Gray and W. Compston, Nature 251:495 (1974)
7.21   T. Lee, D.A. Papanastassiou and G.J. Wasserburg, Astrophysical J. 211:L107 (1977)

sence of ^{26}Al (half-life 7.3×10^5 years) in the early solar system. These data indicated that the initial ^{26}Al/^{27}Al ratio was $(5.1 \pm 0.6) \times 10^{-5}$ (atom/atom).

In 1977, Clayton and Mayeda [7.22] reported that two Ca-Al rich inclusions from the Allende meteorite underwent large mass-fractionation of oxygen isotopes subsequent to incorporation of the nucleosynthetic ^{16}O-anomaly found in other Allende inclusions. Wasserburg et al. [7.23] found, however, that Mg in these inclusions showed *negative* ^{26}Mg which appeared to require the presence of nuclear effects in magnesium distinct from ^{26}Al decay. They noted that the processes responsible for the O and Mg nuclear effects and the astrophysical site of their occurrence remained undefined.

In 1978, Lee et al. [7.24] found isotopic anomalies in calcium in two Ca-Al rich inclusions of the Allende meteorite. The calcium data, when corrected for mass fractionation by using ^{40}Ca/^{44}Ca as a standard, showed non-linear isotopic effects in ^{48}Ca of +13.5 per mil and in ^{42}Ca of +1.7 per mil for one sample. The second sample showed a ^{48}Ca depletion of -2.9 per mil, but all other isotopes were normal. Lee et al. [7.24] concluded that these results required the addition and preservation in the solar system of components from diverse nucleosynthetic sources and, assuming formation of these inclusions by condensation from a gaseous part of the solar nebula, this implied isotopic heterogeneity on a scale of 10 to 10^2 km within the nebula.

In 1978, McCulloch and Wasserburg [7.25] found isotopic anomalies for barium and neodymium in two inclusions from the Allende meteorite. These inclusions were typical Ca-Al-rich objects associated with early condensates from the solar nebula. Sample C1 showed a depletion of only ^{135}Ba of 2 parts in 10^4 and normal neodymium. Sample EK1-4-1 showed large positive excesses in the unshielded isotopes ^{135}Ba and ^{137}Ba of 13.4 and 12.3 parts in 10^4, respectively. They reported that these observations, in conjunction with the presence of ^{26}Al in the early solar system, could be interpreted as the result of a nearby supernova explosion which produced elements over a wide mass range and injected them into the early solar nebula shortly before condensation. The neodymium isotopic composition in EK1-4-1 was highly aberrant in at least five isotopes (see also Lugmair [7.26]).

In 1978, McCulloch et al. [7.27] reported on the samarium isotopic anomalies which they found in the Allende inclusions EK1-4-1 and C1. They noted that their earlier discussion of Ba, Nd and Sm data for Ek1-4-1 and the Sm data for the Allende inclusion C1 had been based on the premise of addition of exotic materials to materials with a normal solar-system isotopic composition, but the presence of negative anomalies in other elements (for example, ^{135}Ba and ^{48}Ca in C1) suggested that this may not have been the case and materials which received these additions may have been depleted in some isotopes. Lugmair [7.26] pointed out, however, that the isotopic excesses obtained for samarium and neodymium from EK1-4-1 appeared to be consistent with average solar system *r*-process abundance. He showed that the observed REE anomalies could be explained as due to a normal *r*-process addition or *s*-process depletion.

7.22   R.N. Clayton and T.K. Mayeda, Geophys. Res. Lett. 4:295 (1977)

7.23   G.J. Wasserburg, T. Lee and D.A. Papanastassiou, Geophys. Res. Lett. 4:299 (1977)

7.24   T. Lee, D.A. Papanastassiou and G.J. Wasserburg, Astrophys. J. 220:L21 (1978)

7.25   M.T. McCulloch and G.J. Wasserburg, Astrophys. J. 220:L15 (1978)

7.26   G.W. Lugmair, Geological Survey Open-File Report 78-701:262 (1978)

7.27   M. McCulloch, G.J. Wasserburg, and D.A. Papanastassiou, Geological Survey Open-File Report 78-701:282 (1978)

Also in 1978, Kelly and Wasserburg [7.28] found the ^{107}Ag/^{109}Ag ratio in the Santa Clara iron meteorite to be 4 percent greater than the terrestrial value and they estimated the initial ^{107}Pd/^{110}Pd ratio in the solar system material to be $\geqslant 2 \times 10^{-5}$. They reported that although the data were strongly suggestive of an excess in ^{107}Ag due to ^{107}Pd (half-life 6.5×10^6 years), the evidence was not definitive and the question, of whether the last nucleosynthesis event was due to a supernova or the result of solar system processes in the general sense of the 1962 hypothesis of Fowler, Greenstein and Hoyle [7.6], was not resolved.

It is perhaps important to note here that Black [7.29], in 1972, announced that he discovered in carbonaceous chondrites a strange neon component, *Neon-E*, which appeared to be the product of a stellar nucleosynthesis. In 1974, Eberhardt [7.30] reported that he was able to separate a Neon-E rich phase from the Orgueil carbonaceous chondrite, and gave the following best estimate for the isotopic composition: $0.45 \leqslant {}^{20}$Ne/^{22}Ne < 1.30 and ^{21}Ne/^{22}Ne < 0.015. In 1978, Srinivasan and Anders [7.31] reported that neon released at 1,200 to 1,600 °C from a severely etched mineral fraction 1C10 of the Murchison meteorite was highly enriched in ^{22}Ne and, whatever the nature of this anomalous neon component, it showed up in the same fractions where the anomalous krypton and xenon were found.

Also in 1972, Manuel et al. [7.32] noted that in xenon fractions released from carbonaceous chondrites at 600 to 1,000 °C an enrichment of the light isotopes ^{124}Xe and ^{126}Xe always correlated with an enrichment of the heavy isotopes ^{134}Xe and ^{136}Xe. They suggested that these xenon fractions may contain material that has been added to our solar system from a nearby supernova, since these are the isotopes expected to be produced in supernova explosions: the $^{124, 126}$Xe isotopes by the *p*-process and the $^{134, 136}$Xe by the *r*-process.

In 1979, Heydegger et al. [7.33] reported that some inclusion materials from the Allende meteorite had a statistically significant enhancement of the order of one part per mil in the ^{50}Ti/^{49}Ti ratio, probably due to a nucleogenic anomaly in ^{50}Ti abundance. Anomalies have been also reported for a number of other elements [7.34–7.37]. The subject of isotopic anomalies has been reviewed by Black [7.38], Reynolds [7.39], Clayton [7.40], Podosek [7.41], Lee [7.42], and Begemann [7.43].

7.28   W.R. Kelly and G.J. Wasserburg, Geophys. Res. Lett. 5:1079 (1978)
7.29   D.C. Black, Geochim. Cosmochim. Acta 36:347 (1972)
7.30   P. Eberhardt, Earth Planet. Sci. Lett. 24:182 (1974)
7.31   B. Srinivasan and E. Anders, Science 201:51 (1978)
7.32   O.K. Manuel, E.W. Hennecke and D.D. Sabu, Nature Phys. Sci. 240:99 (1972)
7.33   H.R. Heydegger, J.J. Foster and W. Compston, Nature 278:704 (1979)
7.34   S. Jovanovic and G.W. Reed, Jr., Earth Planet. Sci. Lett. 31:95 (1976)
7.35   J.W. Arden, Nature 269:788 (1977)
7.36   R.V. Ballad, L.L. Oliver, R.G. Downing and O.K. Manuel, Nature 277:615 (1979)
7.37   S. Yanagita and R. Gensho, Geochem. J. 11:41 (1977)
7.38   D.C. Black, The Origin of the Solar System, edited by S.F. Dermott, John Wiley & Sons, New York, 1978, p. 583
7.39   J.H. Reynolds, Cosmochemistry, edited by W.O. Milligan, Proceedings of the Robert A. Welch Foundation Conferences on Chemical Research. XXI. Houston, Texas, 1978, p. 201
7.40   R.N. Clayton, Ann. Rev. Nucl. Part. Sci. 28:501 (1978)
7.41   F.A. Podosek, Ann. Rev. Astron. Astrophys. 16:293 (1978)
7.42   T. Lee, Rev. Geophys. Space Phys. 17:1591 (1979)
7.43   F. Begemann, Rep. Prog. Phys. 43:1309 (1980)

The opinions of the above-mentioned authors of review articles are unanimous in that the isotopic anomalies observed in meteorites are attributable to the fact that the solar system was not originally homogeneous, in the sense that the products of the various galactic nucleosynthesis processes that created its substance were never completely mixed. If this conclusion is accepted to be correct, however, great difficulties are encountered in correlating the anomalies observed in one element to the next and the experimental data can not be explained in a straightforward manner. In order to overcome these difficulties, the following possibility will be considered next: *The nuclear processes, which were responsible for the production of lithium, beryllium and boron, were also instrumental in the creation of the isotopic anomalies observed in meteorites.*

7.3. A Unified Theory of Isotopic Anomalies

Table 7.1 shows the values of partial cross sections (in millibarns) for production of various nuclides from collisions of six major cosmic-ray nuclides with hydrogen at energies greater than 2.3 GeV/nucleon, which were compiled by Shapiro [7.44]. We shall use these values to calculate the extent to which the isotopic compositions of oxygen and magnesium in meteorites may have been altered by the cosmic ray irradiation.

Let us first consider the following relationship:

$$\frac{(^{17}O)_c}{(^7Li)_c} = \frac{N_3 \cdot \sigma_3^* + N_4 \cdot \sigma_4^* + N_5 \cdot \sigma_5^* + N_6 \cdot \sigma_6^*}{N_1 \cdot \sigma_1 + N_2 \cdot \sigma_2 + N_3 \cdot \sigma_3 + N_4 \cdot \sigma_4 + N_5 \cdot \sigma_5 + N_6 \cdot \sigma_6} \quad (7.1)$$

where the N's are the cosmic abundances; the σ's are the partial cross sections for production of ^7Li; the σ^*'s are the partial cross sections for production of ^{17}O; the subscripts 1, 2, 3, 4, 5, and 6 refer to the six major cosmic-ray nuclides ^{12}C, ^{16}O, ^{20}Ne, ^{24}Mg, ^{28}Si and ^{56}Fe, respectively; and the subscript c refers to the number of atoms, which were produced by the cosmic rays.

If it is assumed that all the ^7Li atoms in the solar system were produced by the cosmic rays, the value of $(^{17}O)_c$ can be calculated from the cosmic abundance value of ^7Li by the use of equation (7.1). We shall use the value of 92.6 for the cosmic abundance of ^7Li (see the Suess-Urey cosmic abundance table, Appendix II). Introducing this value and the values of N's, σ's, and σ^*'s shown in the Suess-Urey table and Table 7.1 into equation (7.1), we have a value of $(^{17}O)_c = 47.8$. The cosmic abundance of ^{17}O is 8,000 and hence as much as

$$\frac{(47.8)\,(1,000)}{(8,000)} = 6.0 \text{ per mil}$$

of the ^{17}O in the matter within the solar system may be cosmic-ray-produced.

The content of $(^{17}O)_c$ in the matter within the solar system may be considerably higher than this, since it is quite possible that planetesimals, in which the ratio of the oxygen atoms to the atoms of magnesium, silicon, iron, etc. was much smaller than the

---

7.44 M.M. Shapiro, 12th Int. Conf. Cosmic Rays, Hobart, Tasmania, Australia, 16–25 August 1971 (Rapporteur Paper), p. 422

Table 7.1. Partial cross sections (in millibarns) for production of various nuclides from collisions of six major cosmic-ray nuclides with hydrogen; $E \geqslant 2.3$ GeV/nucleon (according to Shapiro [7.44])

Product	Target[a]					
	^{12}C	^{16}O	^{20}Ne	^{24}Mg	^{28}Si	^{56}Fe
^{6}Li	7[b]	14	12	13	13	30
^{7}Li	6	14	11	11	11	20
^{7}Be	10	11	10	10	10	8.5
^{9}Be	6	3.7	3	3	3	5
^{10}Be	3.5	1.0	1.9	1.9	1.9	4
^{10}B	14	12	9	8	7	7
^{11}B	51	25	18	15	12	9
^{12}C		24	18	13	10	7
^{13}C		20	14	10	8	5
^{14}N		26	18	13	10	6
^{15}N		50	23	17	13	6
^{16}O			24	18	13	6
^{17}O			25	19	14	6
^{18}O			23	16	12	6
^{19}F			45	19	14	7
^{20} ^{21} ^{22}Ne				69	55	21
^{23}Na				51	23	9
^{24} ^{25} ^{26}Mg					77	25
^{27}Al					52	10
σ_i	205[c]	260	315	355	400	676

[a] A cosmic-ray nucleus which collides with interstellar hydrogen

[b] Values in italics refer to cross sections based primarily upon experimental information

[c] The quantity σ_i is the total inelastic cross section (including, e.g., for helium production for which values are not given here)

ratio corresponding to their cosmic abundance values, were irradiated with high energy charged particles. If it is assumed that the chemical composition of the planetesimals was approximately the same as that of an average silicate meteorite, the ratio of the oxygen atoms to the sum of the number of atoms of Si, Al, Fe, Ca, Mg and Na is about 1.4, while the same ratio for the cosmic abundance values of these elements is 7.9 (atom/atom). It follows then that as much as

$$\frac{(6.0)\,(7.9)}{(1.4)} = 34 \text{ per mil}$$

of the ^{17}O in the meteorites may be cosmic-ray-produced. This means that ^{17}O may be *deficient* as much as 34 per mil relative to its normal abundance in some meteorites, which have not been affected by the cosmic ray irradiation.

The cosmic abundance of ^{18}O is about 5.5 times that of ^{17}O and the formation cross sections of ^{18}O from ^{20}Ne, ^{24}Mg, ^{28}Si, and ^{56}Fe are roughly the same as those of ^{17}O.

Thus, ^{18}O may be *deficient* in solid bodies which have not been affected by the cosmic ray irradiation by as much as $34/5.5 = 6.2$ per mil.

Since individual cross section values for the production of magnesium isotopes from ^{28}Si were not reported by Shapiro [7.44], we shall assume here that the value for ^{26}Mg is about one third of the value of 77 millibarns for $^{24, 25, 26}Mg$ or about 26 millibarns. The cosmic abundance of ^{26}Mg is 1.00×10^5, while that of ^{18}O is 4.36×10^4. It follows then that ^{26}Mg may be *deficient* in solid bodies which have not been affected by the cosmic ray irradiation by as much as

$$\frac{(6.2)\,(26)\,(4.36 \times 10^4)}{(16)\,(1.00 \times 10^5)} = 4.4 \text{ per mil}$$

relative to its normal cosmic abundance value.

The extents to which the abundances of ^{17}O, ^{18}O, and ^{26}Mg may be depleted in solid bodies which have not been affected by the cosmic ray irradiation were similarly caculated from the cosmic abundance values of lithium, beryllium, and boron and the results are compared in Tables 7.2 and 7.3 with the experimental data obtained by Clayton et al. [7.22, 7.45] and Wasserburg et al. [7.23] for the oxygen and the magnesium, respectively, from the mineral inclusions of the Allende meteorite. The extents to which ^{17}O and ^{18}O are depleted in the Allende inclusions relative to their normal abundances are shown here as $(\delta^{17}O)_c$ and $(\delta^{18}O)_c$, respectively. These values were calculated from the values of $\delta^{17}O$ and $\delta^{18}O$ reported by Clayton et al. [7.22, 7.45] as it is explained below.

The relationship between the isotopic compositions of oxygen in meteoritic and terrestrial samples may be written in the form [7.46]

$$M^{17} = T^{17}\,(1 + \mu) + c_{17}$$

and (7.2)

$$M^{18} = T^{18}\,(1 + 2\mu) + c_{18}$$

where M^i and T^i are the $^iO/^{16}O$ ratios in the meteoritic and terrestrial samples, respectively; μ is a mass fractionation factor; and c_i is the contribution from the cosmic-ray-produced isotope at mass number i.

The oxygen isotope data are usually expressed in the form

$$\delta^iO = \left\{ (^iO/^{16}O)_M / (^iO/^{16}O)_{SMOW} - 1 \right\} \cdot (1{,}000) \text{ (per mil)} \qquad (7.3)$$

where the subscripts M and SMOW refer to the $^iO/^{16}O$ ratio in the sample and in the standard mean ocean water. Assuming that the values of $(^iO/^{16}O)_{SMOW}$ to be equal to the values of T^i, we have the following relationships from equations (7.2) and (7.3),

7.45   R.N. Clayton, N. Onuma, L. Grossman and T.K. Mayeda, Earth Planet. Sci. Lett. 34:209 (1977)
7.46   P.K. Kuroda, Geochem. J. 9:51 (1975)

Table 7.2. Isotopic anomalies observed in the oxygen from the mineral inclusions of the Allende meteorite[a]

Sample		$(\delta^{17}O)_c$ (per mil)	$(\delta^{18}O)_c$ (per mil)
(I)	*Experimental values*		
(1)	EK-1-4-1Me1	− 3.7	−0.6
(2)	C1S2	−11.4	−1.9
(3)	C1S3	−12.7	−2.1
(4)	EK1-4-1Cpx	−20.5	−3.3
(5)	IK1-4-1Sp + Cpx	−22.6	−3.7
(6)	EK1-4-1Sp	−23.4	−3.8
(7)	A13S4	−27.8	−4.5
Average		−17.4	−2.8
(II)	*Calculated values*		
(1)	From $H = 0.81$ for ^{9}Be[b]	− 1.0	−0.2
(2)	From $H = 7.4$ for ^{6}Li[c]	− 2.5	−0.5
(3)	From $H = 26$ for ^{10}B[e]	−10.5	−1.9
(4)	From $H = 45.8$ for ^{7}Li[d]	−16.9	−3.1
(5)	From $H = 114$ for ^{11}B[e]	−20.9	−3.5
(6)	From $H = 20$ for ^{9}Be[c]	−24.2	−4.4
(7)	From $H = 92.6$ for ^{7}Li[c]	−34.2	−6.2
Average		−15.8	−2.8

(a)   The original experimental data were reported by Clayton and Mayeda [7.22] and Clayton et al. [7.45]

(b)   Based on the cosmic abundance value of ^{9}Be recommended by Mason (1971) [7.47] and others

(c)   Suess and Urey [7.48]

(d)   Based on the value of 49.5 for lithium recommended by Mason (1971) [7.47] and others

(e)   Based on the value of 140 for boron recommended by Mason [7.47] and others

$$(\delta^i O) \quad = (\delta^i O)_m + (\delta^i O)_c$$
$$(\delta^{17}O)_m = \mu \cdot (1{,}000)$$
$$(\delta^{18}O)_m = \mu \cdot (2{,}000) \tag{7.4}$$
$$(\delta^{17}O)_c = \frac{c_{17}}{T^{17}} \cdot (1{,}000)$$

and

$$(\delta^{18}O)_c = \frac{c_{18}}{T^{18}} \cdot (1{,}000)$$

7.47   B. Mason, Handbook of Elementary Abundances in Meteorites, Gordon and Breach, New York, 1971, 555 pp.

Table 7.3. Isotopic anomalies observed in the magnesium from the mineral inclusions of the Allende meteorite[a]

Sample		$(\delta^{26}Mg)_C$ (per mil)
(I)	*Experimental values*	
(1)	C1S1	-1.5 ± 0.2
(2)	C1 Spinel	-1.6 ± 0.2
(3)	C1S1	-1.7 ± 0.3
(4)	C1 Spinel	-1.9 ± 0.8
(5)	C1S1	-2.0 ± 0.2
(6)	EK1-4-1 Spinel B	-2.9 ± 0.8
(7)	EK1-4-1 Fassaite A	-3.6 ± 0.3
(8)	EK1-4-1 Spinel A	-3.9 ± 0.5
Average		-2.4
(II)	*Calculated values*	
(1)	From $H = 0.81$ for ^{9}Be[b]	-0.1
(2)	From $H = 7.4$ for ^{6}Li[c]	-0.4
(3)	From $H = 26$ for ^{10}B[e]	-1.4
(4)	From $H = 45.8$ for ^{7}Li[d]	-2.2
(5)	From $H = 114$ for ^{11}B[e]	-2.5
(6)	From $H = 20$ for ^{9}Be[c]	-3.1
(7)	From $H = 92.6$ for ^{7}Li[c]	-4.4
Average		-2.0

[a]   The original experimental data were reported by Wasserburg et al. [7.23] in 1977; (b), (c), (d), and (e), see the footnotes to Table 7.2

where $T^{17} = 0.0003910$ and $T^{18} = 0.0020550$. The ratio of c_{17} to c_{18} was assumed to be equal to the ratio of partial cross sections for production of ^{17}O and ^{18}O from ^{28}Si, which is $14/12 = 1.17$ (see Table 7.1). Fig. 7.1 illustrates the relationship between (δ^iO), $(\delta^iO)_m$ and $(\delta^iO)_c$.

The relationship between the isotopic compositions of magnesium in meteoritic and terrestrial samples may be written similarly in the form

$$M^{25} = T^{25}(1 + \mu) + c_{25}$$

and (7.5)

$$M^{26} = T^{26}(1 + 2\mu) + c_{26}$$

where M^i and T^i are the $^iMg/^{24}Mg$ ratios in the meteoritic and terrestrial samples, respectively; μ is a mass fractionation factor; and c_i is the contribution from the cosmic-ray-produced isotope at mass number i.

124

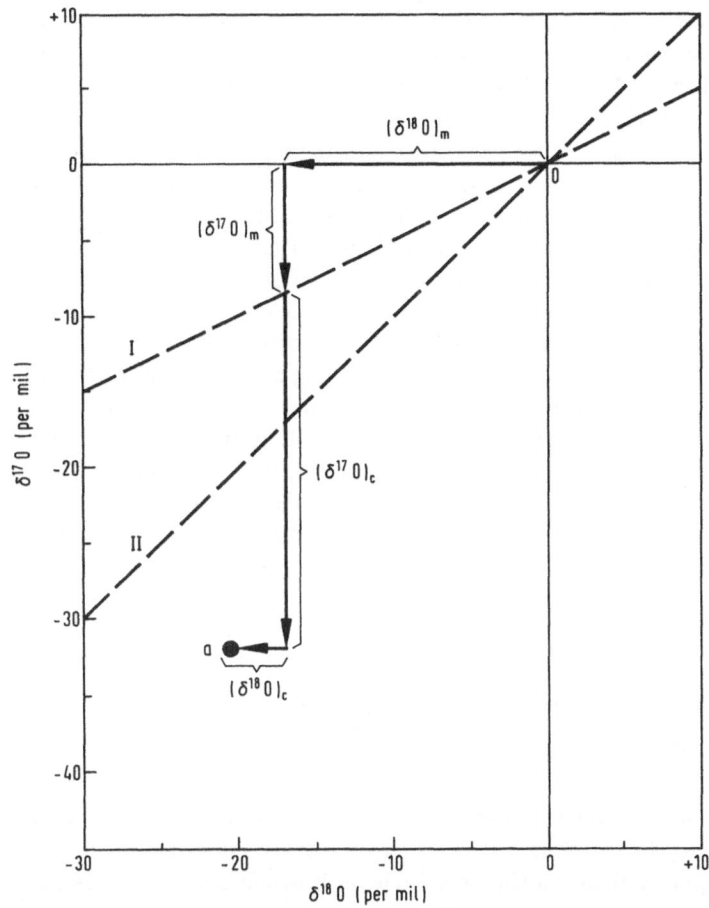

Fig. 7.1. The Puzzle of the Oxygen. The isotopic composition of oxygen in the Allende inclusion EK1-4-1Sp is shown by the point *a*. According to Clayton and Mayeda [7.22], the inclusions underwent large mass-fractionation of oxygen isotopes subsequent to incorporation of the nucleosynthetic ^{16}O-anomaly found in other Allende inclusions. The data can be interpreted in a different manner as due to an alteration of the isotopic composition by a combined effect of mass-fractionation and cosmic-ray irradiation processes. The subscripts *m* and *c* refer to the alterations of the isotopic composition by mass-fractionation and cosmic-ray irradiation processes, respectively. The two dotted lines I and II have the equations: Line I: $\delta^{18}O = 2\delta^{17}O$, Line II: $\delta^{18}O = \delta^{17}O$

Wasserburg et al. [7.23] assumed the value of c_{25} in equation (7.5) to be equal to zero and calculated the value of $\delta^{26}Mg$. This may not necessarily have been a bad assumption, since the abundance of ^{26}Mg is expected to be altered also by the decay of 7.3×10^5-year ^{26}Al, which will be produced from ^{28}Si by the cosmic ray irradiation. Furukawa et al. [7.49] have reported the formation cross section of ^{26}Al from pro-

7.48   H.E. Suess and H.C. Urey, Rev. Mod. Phys. 28:53 (1956)
7.49   M. Furukawa, K. Shizuri, K. Komura, K. Sakamoto and S. Tanaka, Nucl. Phys. A174:539 (1971)

ton bombardment of natural Si at 44 MeV to be 60 millibarns, which is comparable to the value of 77 millibarns for the production of ^{24}Mg, ^{25}Mg, and ^{26}Mg combined (see Table 7.1). The values of δ^{26}Mg reported by Wasserburg et al. [7.23] are shown in Table 7.3 as $(\delta^{26}$Mg$)_c$.

Tables 7.2 and 7.3 show that the extents to which the abundances of ^{17}O, ^{18}O, and ^{26}Mg are depleted in the high temperature mineral inclusions of the Allende meteorite are in general agreements with the values of $(\delta^{17}$O$)_c$, $(\delta^{18}$O$)_c$, and $(\delta^{26}$Mg$)_c$ calculated from the cosmic abundances of lithium, beryllium, and boron. These results seem to indicate that the oxygen and magnesium anomalies are not caused by addition of "exotic" materials to materials with a normal isotopic composition, but they are likely to have originated from the fact that planetesimals within the solar system did not receive a uniform irradiation by charged particles during an early stage of the history of their formation. It thus appears that the Allende inclusions represent the solar system materials, which received the smallest amount of charged particle irradiation during their past history.

If the above conclusion is correct, the abundances of lithium, beryllium, and boron in the mineral inclusions of the Allende meteorite are expected to be almost zero. In this connection, it is interesting to note that Weller et al. [7.50] have recently reported that they found the abundance of boron to be equal to or less than 10^{16} atoms per gram in two Allende fine-grained inclusions. They also noted that results from their particle track method of boron analysis of the inclusions indicated that either all tracks were produced by lithium, and boron was highly depleted, or even if lithium were anomalously absent in the samples, the resulting boron upper limit was only 0.5 p.p.m. B.

It is worthy of note in Tables 7.2 and 7.3 that the values of $(\delta^{17}$O$)_c$, $(\delta^{18}$O$)_c$, and $(\delta^{26}$Mg$)_c$ calculated from the values of H = 0.81, 7.4, and 26 for ^{9}Be, ^{6}Li, and ^{10}B, respectively, are in poor agreement with the values calculated from the abundances of ^{7}Li and ^{11}B and the Suess-Urey value of H = 20 for ^{9}Be. The reason for these discrepancies may lie in the fact that the cross section values for the protons with energies equal to or greater than 2.3 GeV per nucleon shown in Table 7.1 were used for the calculations. Weller et al. [7.50] have recently suggested that a substantial fraction of lithium, beryllium, and boron must have been produced by a large flux of low energy particles. According to Jacobs et al. [7.17], ratios of ^{7}Li/^{6}Li, ^{11}B/^{10}B, and B/Li yields for proton reactions with C, N, and O vary quite considerably at energies in the vicinity of 20 to 25 MeV.

It is interesting to recall that Fowler et al. [7.6] have once considered the possibility that ^{6}Li and ^{10}B were destroyed by the (n,α) reactions induced by thermal neutrons. The cross section value for ^{10}B is 3,837 barns and hence the integrated thermal neutron flux which is required to reduce the cosmic abundance of ^{10}B by a factor of 2 is calculated to be

$$\Phi \cdot t = \frac{(1/2)}{(3,837 \times 10^{-24})} = 1.3 \times 10^{20} \ (\text{n/cm}^2)$$

The results from the studies on the isotopic composition of gadolinium in meteorites seemed to indicate that the neutrons associated with an early irradiation of the solar system materials did not play an important role in altering the isotopic compositions of

---

7.50   M.R. Weller, M. Furst, T.A. Tombrello and D.S. Burnett, Geochim. Cosmochim. Acta 42:999 (1978)

various elements in meteorites, as was discussed in Section 7.1. It is worthy of note here, however, that the effect of secondary neutrons associated with the irradiation of solid bodies with high energy charged particles may have not been negligibly small. The integrated flux of high energy protons, which is required to produce the deficient elements, can be estimated in the following manner: the sum of the cosmic abundance values of ^{12}C, ^{16}O, ^{20}Ne, ^{24}Mg, ^{28}Si, and ^{56}Fe is approximately 3.5×10^7 in the scale of Si equals to 10^6. The values of partial cross sections for production of ^{11}B from the above nuclides are 51, 25, 18, 15, 12, and 9 millibarns, as shown in Table 7.1, with an average value of 22 millibarns. If it is assumed that the partial cross sections for production of ^{11}B from each of the above-mentioned nuclides be equal to 22 millibarns and the cosmic abundance value of H = 114 is used for ^{11}B, we have a value of

$$\Phi \cdot t = \frac{(114)}{(3.5 \times 10)(22 \times 10^{-27})} = 1.5 \times 10^{20} \ (p/cm^2)$$

for the integrated proton flux.

The integrated flux of secondary neutrons must have been at least equal to or somewhat larger than the value of integrated proton flux obtained above. Most of these secondary neutrons were obviously not thermalized before they were absorbed by the matter in the solar system and the effect of these "non-thermal" neutrons is expected to be detected in the form of alterations of the isotopic compositions of various elements in meteorites.

As was mentioned in Section 7.2., McCulloch et al. [7.27] found the isotopic anomalies in samarium from the Allende inclusions. Although they attempted to explain the anomalies on the premise of addition of "exotic" materials to materials with a normal solar-system isotopic composition, the anomalies observed in the samarium from the Allende inclusions can be explained in a more straightforward manner as due to the fact that the Allende inclusions received a smaller integrated flux of "non-thermal" neutrons than the average solar system material and the difference was equivalent to a flux of 3.3×10^{20} (n/cm^2) of 15 KeV neutrons [7.51]. Table 7.4 shows a comparison of the isotopic compositions of samarium from the Allende inclusion EK1-4-1SC and neutron-irradiated normal samarium. Note that the flux of "non-thermal" neutrons responsible for the samarium anomalies observed in the Allende inclusions is of the same order of magnitude as that of the integrated proton flux calculated earlier.

As it has recently been pointed out by Kuroda [7.52], anomalies which have recently been observed in the stable isotopes of O, Ne, Mg, Ca, Ti, Kr, Ag, Te, Xe, Ba, Ce, Nd, Sm, and Gd in meteorites are all related to each other and can be explained as due to a combined effect of (a) mass-fractionation, (b) "non-thermal" neutron capture, and (c) spallation processes which took place prior to and/or during the formation of the solar system, as shown in Table 7.5.

In the isotopic anomalies observed in the elements lighter than iron (Z = 26), the processes (a) and (c) play the dominant roles, whereas in the cases of heavier elements

7.51 P.K. Kuroda, Geochem. J. 13:287 (1979)
7.52 P.K. Kuroda, Presented at the Symposium on Practical Applications of Nuclear and Radiochemistry, Second Chemical Congress of the North American Continent, Las Vegas, Nevada, August 26, 1980

Table 7.4. A comparison of the isotopic compositions of samarium in the Allende inclusion and neutron-irradiated normal samarium

Sample		$\epsilon 144$	$\epsilon 147$	$\epsilon 148$	$\epsilon 149$	$\epsilon 150$	$\epsilon 152$	$\epsilon 154$
(1)	Allende inclusion EK1-4-1Sc[a]	-17.0 $\pm\ 5.2$	$\equiv 0$	-36.2 $\pm\ 2.2$	$-\ 0.7$ $\pm\ 1.5$	-35.1 $\pm\ 2.2$	-9.1 ± 1.4	$\equiv 0$
(2)	Neutron-irradiated normal Sm[b]							
	(a) $\Phi t = 4.5 \times 10^{17}$ (n/cm^2)							
	(thermal)	0.0	$\equiv 0$	0.0	-18.7	$+35.1$	$+\ 1.0$	$\equiv 0$
	(b) $\Phi t = 3 \times 10^{20}$ (n/cm^2)							
	(15 KeV)	$+18.9$	$\equiv 0$	$+29.5$	$-\ 0.3$	$+35.1$	$+\ 7.9$	$\equiv 0$

[a] The experimental data were reported by McCulloch et al. [7.24]. The isotopic composition is expressed here in terms of ϵ_i defined as

$$\epsilon_i = \ [(^i Sm/^{154} Sm)_C/(^i Sm/^{154} Sm)_N - 1]\ \cdot 10^4\,,$$

where the subscript N refers to the normal samarium and the subscript C to the ratios measured and normalized for mass-fractionation assuming the variation in the $^{147} Sm/^{154} Sm$ ratio to be entirely due to mass-fractionation

[b] A mixture of Sm, Eu, and Nd in the ratio of cosmic abundances of these elements given by Burbidge et al. [7.5] was assumed to have been irradiated with neutrons

Table 7.5. The relative importance of the processes of mass-fractionation, neutron-capture and spallation responsible for the isotopic anomalies observed in various elements[*]

Element	Atomic number (Z)	Atomic weight (A)	Number of stable isotopes	(a) Mass-fractionation	(b) Neutron-capture	(c) Spallation
Oxygen	8	15.9994	3	yes	no	yes
Neon	10	20.179	3	yes	no	yes
Magnesium	12	24.305	3	yes	no	yes
Calcium	20	40.08	6	yes	no	yes
Titanium	22	47.90	5	yes	no	yes
Krypton	36	83.80	6	yes	yes	yes
Xenon	54	131.30	9	yes	yes	yes
Barium	56	137.34	7	yes	yes	no
Cerium	58	140.12	4	yes	yes	no
Neodymium	60	144.24	7	yes	yes	no
Samarium	62	150.4	7	yes	yes	no
Gadolinium	64	157.25	7	yes	yes	no

[*] The word "yes" means here that the process plays a major role and the word "no" means that it plays a minor role in producing the isotopic anomalies

the anomalies can be attributed to a combined effect of (a) and (b). This is to be expected, since the neutron-capture cross-sections of the isotopes of light elements are generally very small, whereas the cross-section values for many of the isotopes of heavy elements are greater than one barn even at the neutron energies ranging from 10 to 15 KeV. The isotopic anomalies observed in krypton and xenon can be attributed to a combined effect of all three processes (a), (b), and (c). This is because they are gaseous elements and their abundances in solid bodies are much smaller than those of neighboring elements.

Radionuclides such as 16-million year ^{129}I, 6.5-million year ^{107}Pd and 0.73-million year ^{26}Al were produced by either the spallation or neutron-capture processes which took place prior to and/or during the formation of the solar system. Thus it appears that these relatively short-lived "extinct" radionuclides are not the products of the galactic nucleosynthesis processes, but they were produced sometime during the period between the formation of the elements in stars and the condensation of the solar nebula. The apparent conflict between the time scales inferred from the cosmochronometers ^{129}I, ^{244}Pu and ^{26}Al, which has been pointed out by Lee et al. [7.42, 7.53], can be eliminated if the existing experimental data are interpreted in this manner.

A detailed discussion on each of the isotopic anomalies observed in meteorites is beyond the scope of this monograph. Brief discussions on a number of important cases will be presented in the second half of this chapter.

7.4. Neon

Neon has three stable isotopes: ^{20}Ne, ^{21}Ne and ^{22}Ne. The case of neon is quite analogous to the case of oxygen and magnesium discussed in the previous section. The relationship between the isotopic composition of neon in meteoritic and terrestrial samples can be written in the form [7.54]

$$M^{21} = T^{21} (1 + \mu) + c_{21}$$

and (7.6)

$$M^{22} = T^{22} (1 + 2\mu) + c_{22}$$

where M^i and T^i refer to the ^{i}Ne/^{20}Ne ratios in meteorites and in the atmosphere, respectively; μ is a mass-fractionation factor; c_i is the contribution from the cosmic-ray-produced isotope at mass number i.

A great significance has been attached to the suggestion made by Black [7.29] in 1972 that carbonaceous chondrites contain a strange neon component: Neon-E, which appears to be the product of a stellar nucleosynthesis (see Section 7.2.). Eberhardt [7.30] reported that he was able to separate a Neon-E rich phase from the Orgueil carbonaceous chondrite and gave the following best estimate for the isotopic composition:

7.53 T. Lee, D.N. Schramm, J.P. Wefel and J.B. Blake, Geological Survey Open-File Report 78-701: 246 (1978)

7.54 P.K. Kuroda, Geochem. J. 9:51 (1975)

$0.45 \leqslant {}^{20}\mathrm{Ne}/{}^{22}\mathrm{Ne} < 1.30$ and ${}^{21}\mathrm{Ne}/{}^{22}\mathrm{Ne} < 0.015$. Neon-$E$ is abnormally rich in ${}^{22}\mathrm{Ne}$ and it appears that we are dealing here with the decay product of spallation-produced ${}^{22}\mathrm{Na}$:

$$ {}^{22}\mathrm{Na} \xrightarrow[\text{2.601-year}]{\beta^+} {}^{22}\mathrm{Ne}\ (\text{stable}) $$

7.5. Argon

Argon has three stable isotopes: ${}^{36}\mathrm{Ar}$, ${}^{38}\mathrm{Ar}$ and ${}^{40}\mathrm{Ar}$. The case of argon is somewhat complicated since the radioactive decay of ${}^{40}\mathrm{K}$ contributes to the abundance of ${}^{40}\mathrm{Ar}$. Moreover, the abundance of ${}^{36}\mathrm{Ar}$ is sometimes enhanced by the reactions

$$ {}^{35}\mathrm{Cl}(\mathrm{n},\gamma){}^{36}\mathrm{Cl} \xrightarrow[\text{3.01} \times 10^5\ \text{y}]{\epsilon,\ \beta^+} {}^{36}\mathrm{Ar}\ (\text{stable}) $$

The thermal neutron-capture cross-section of ${}^{35}\mathrm{Cl}$ is 43 barns, while that of ${}^{37}\mathrm{Cl}$ is about 0.4 barns. The presence of about 0.4×10^{-8} cc STP/g of excess ${}^{36}\mathrm{Ar}$ in the Allende meteorite was reported by Manuel et al. [7.55] and they attributed this to the neutron-capture reactions on ${}^{35}\mathrm{Cl}$.

The relationship between the isotopic compositions of argon in meteoritic and terrestrial samples can be written in the form [7.54]

$$ (1+r_{40})\, M^{36} = T^{36}\, (1+\mu) + c_{36} + n_{36} $$

and

$$ (1+r_{40})\, M^{38} = T^{38}\, (1+2\mu) + c_{38} $$

(7.7)

where M^i and T^i refer to the ${}^i\mathrm{Ar}/{}^{40}\mathrm{Ar}$ ratios in meteorites and in the atmosphere, respectively; μ is a mass-fractionation factor; c_i is the contribution from the cosmic-ray-produced isotope at mass number i; n_{36} is the contribution from the neutron-capture reactions on ${}^{36}\mathrm{Cl}$; and r_{40} is the contribution from the decay of ${}^{40}\mathrm{K}$.

7.6. Krypton

Krypton has six stable isotopes: ${}^{78}\mathrm{Kr}$, ${}^{80}\mathrm{Kr}$, ${}^{82}\mathrm{Kr}$, ${}^{83}\mathrm{Kr}$, ${}^{84}\mathrm{Kr}$ and ${}^{86}\mathrm{Kr}$. The relationship between the isotopic compositions of the meteoritic and terrestrial (atmospheric) krypton can be written in the form [7.54]

$$
\begin{aligned}
(1+n_{86}+c_{86}) \cdot M^{78} &= A^{78} \cdot (1+8\mu) + n_{78} + c_{78} \\
(1+n_{86}+c_{86}) \cdot M^{80} &= A^{80} \cdot (1+6\mu) + n_{80} + c_{80} \\
(1+n_{86}+c_{86}) \cdot M^{82} &= A^{82} \cdot (1+4\mu) + n_{82} + c_{82} \\
(1+n_{86}+c_{86}) \cdot M^{83} &= A^{83} \cdot (1+3\mu) + n_{83} + c_{83} \\
(1+n_{86}+c_{86}) \cdot M^{84} &= A^{84} \cdot (1+2\mu) + n_{84} + c_{84}
\end{aligned}
$$

(7.8)

7.55   O.K. Manuel, R.J. Wright, D.K. Miller and P.K. Kuroda, Geochim. Cosmochim. Acta 36:961 (1972)

where M^i and A^i are the ^{i}Kr/^{86}Kr ratios in the meteorite and in the earth's atmosphere, respectively; μ is a mass-fractionation factor; n_i is the contribution from the neutron-capture processes at mass number i; and c_i is the contribution from the cosmic-ray-produced isotopes at mass number i.

The values of n_{86} and c_{86} in equation (7.8) can be considered as negligibly small and hence the relationship between the meteoritic and atmospheric krypton can be approximated by

$$
\begin{aligned}
M^{78} &= A^{78} \, (1 + 8\mu) + n_{78} + c_{78} \\
M^{80} &= A^{80} \, (1 + 6\mu) + n_{80} + c_{80} \\
M^{82} &= A^{82} \, (1 + 4\mu) + n_{82} + c_{82} \\
M^{83} &= A^{83} \, (1 + 3\mu) + n_{83} + c_{83} \\
M^{84} &= A^{84} \, (1 + 2\mu) + n_{84} + c_{84}
\end{aligned}
\tag{7.9}
$$

In the case of meteorites with very high gas contents, the mass-fractionation process plays a dominant role and contributions from the neutron-capture and cosmic-ray spallation processes (n_i's and c_i's, respectively) are negligibly small. Hence the relationship between the isotopic compositions of meteoritic krypton and atmospheric krypton can be written in the form

$$
\begin{aligned}
M^{78} &= A^{78} \, (1 + 8\mu) \\
M^{80} &= A^{80} \, (1 + 6\mu) \\
M^{82} &= A^{82} \, (1 + 4\mu) \\
M^{83} &= A^{83} \, (1 + 3\mu) \\
M^{84} &= A^{84} \, (1 + 2\mu)
\end{aligned}
\tag{7.10}
$$

Table 7.6 shows that the meteorites Murray, Novo Urei and Lancé studied by Marti [7.56] belong to this group. The experimental data are expressed here in the form

$$
\delta i = \left[(^iKr/^{86}Kr)_M / (^iKr/^{86}Kr)_A - 1 \right] \cdot (1{,}000) \ \text{(per mil)}
\tag{7.11}
$$

Table 7.6. Isotopic compositions of krypton in meteorites with very high gas concentrations

Meteorite	$\delta 78$	$\delta 80$	$\delta 82$	$\delta 83$	$\delta 84$	$\delta 86$	^{86}Kr content $(10^{-12}$ cc STP/g)
(1) Murray	−37.6 ± 12.5	−18.5 ± 11.6	−20.7 ± 7.6	−15.2 ± 6.1	−13.1 ± 6.1	≡ 0	3,780
(2) Novo Urei	−32.6 ± 15.0	−23.1 ± 9.3	−16.2 ± 7.6	−13.6 ± 6.1	−13.1 ± 6.1	≡ 0	8,050
(3) Lance	−42.6 ± 15.0	−16.2 ± 11.6	−20.7 ± 9.1	−13.6 ± 7.6	−13.1 ± 9.2	≡ 0	1,950
(4) Mass-fractionated atmospheric Kr	−40.0	−30.0	−20.0	−15.0	−10.0	≡ 0	−

7.56   K. Marti, Earth Planet. Sci. Lett. 3:243 (1967)

Table 7.7. Isotopic composition of krypton in meteorites with very low gas contents

Meteorite	$\delta 78$	$\delta 80$	$\delta 82$	$\delta 83$	$\delta 84$	$\delta 86$	^{86}Kr content $(10^{-12}$ cc STP/g)
(1) Petersburg	–	+1976 $\pm$ 138	+569 $\pm$ 36	+680 $\pm$ 41	+65.3 $\pm$18.0	$\equiv 0$	18.3
(2) Bruderheim	–	+ 408 $\pm$ 62	+142 $\pm$ 27	+152 $\pm$ 32	+51.0 $\pm$18.3	$\equiv 0$	54.3
(3) Pasamonte	–	+ 217 $\pm$ 31	+ 49.5 $\pm$ 16.5	+ 61.7 $\pm$ 12.0	+ 4.3 $\pm$ 4.6	$\equiv 0$	67.6

where the subscripts M and A refer to the ^{i}Kr/^{86}Kr ratios in the meteorite and in the earth's atmosphere and i is the mass numbers.

In the case of meteorites with very low gas contents, the cosmic-ray spallation process plays a dominant role and contributions from the mass-fractionation and neutron-capture processes (μ and n_i's, respectively) are negligibly small. Hence the relationship between the isotopic compositions of meteoritic and atmospheric krypton can be written in the form

$$
\begin{aligned}
M^{78} &= A^{78} + c_{78} \\
M^{80} &= A^{80} + c_{80} \\
M^{82} &= A^{82} + c_{82} \\
M^{83} &= A^{83} + c_{83} \\
M^{84} &= A^{84} + c_{84}
\end{aligned}
\tag{7.12}
$$

Table 7.7 shows that the meteorites Petersburg, Pasamonte, and Bruderheim studied by Bogard [7.57] and Beck et al. [7.58] belong to this group.

In the case of meteorites with intermediate gas contents, the neutron-capture process plays a dominant role and contributions from the mass-fractionation and cosmic-ray spallation processes (μ and c_i's, respectively) are negligibly small. Hence the relationship between the isotopic compositions of meteoritic and atmospheric krypton can be written in the form

$$
\begin{aligned}
M^{78} &= A^{78} + n_{78} \\
M^{80} &= A^{80} + n_{80} \\
M^{82} &= A^{82} + n_{82} \\
M^{83} &= A^{83} + n_{83} \\
M^{84} &= A^{84} + n_{84}
\end{aligned}
\tag{7.13}
$$

Table 7.8 shows that the meteorites Mezö-Madaras, Parnallee and Allende studied by Eugster et al. [7.59], Lewis et al. [7.60] and Manuel et al. [7.55] belong to this group.

7.57 D.D. Bogard, J. Geophys. Res. 72:1299 (1967)
7.58 J.N. Beck, D.K. Miller, D.W. Efurd, R.H. Thompson, M.A. Reynolds, K. Sakamoto and P.K. Kuroda (unpublished manuscript, 1972)
7.59 O. Eugster, P. Eberhardt and J. Geiss, Earth Planet. Sci. Lett. 3:385 (1967)
7.60 R.S. Lewis, B. Srinivasan and E. Anders, Science 190:1251 (1975)

Table 7.8. Isotopic composition of krypton in meteorites with intermediate gas contents

Meteorite	$\delta 78$	$\delta 80$	$\delta 82$	$\delta 83$	$\delta 84$	$\delta 86$	^{86}Kr content $(10^{-12}$ cc STP/g)
(1) Mezö-Madaras	−10.0 ± 25.0	+363 ± 23	+13.8 ± 13.6	−12.6 ± 9.1	−18.3 ± 9.2	≡ 0	1020
(2) Parnallee	−15.5 ± 15.0	+500 ± 15	+29.9 ± 6.0	− 9.1 ± 4.5	−16.2 ± 6.1	≡ 0	350
(3) Allende	− 7.5 ± 63.2	+313 ± 4	+ 6.6 ± 2.0	−12.1 ± 2.4	−15.3 ± 1.1	≡ 0	811
(4) Allende	−	+273 ± 15	+ 4.3 ± 6.0	−15.0 ± 7.6	− 8.8 ± 3.7	≡ 0	928
(5) Neutron-irradiated atmospheric Kr: $\Phi t = 5 \times 10^{22}$ (n/cm²) (15 KeV)[a]	− 5.0	+129	+21.5	−11.5	− 0.8	≡ 0	−

[a] A mixture of Se, Br and the atmospheric Xe in the ratio of cosmic abundances of these elements given by Burbidge et al. [7.5] was assumed to have been exposed to a flux of 15 KeV neutrons. The total neutron flux was assumed to have been such that the abundance of ^{86}Kr became depleted by 1.2 parts per 10^4

The case of krypton anomalies had always been considered as one of the most ambiguous and baffling of all the anomalies observed in meteorites. It now appears, however, that the anomalies found in krypton can be explained quite satisfactorily as due to a combined effect of (a) mass-fractionation, (b) neutron-capture and (c) cosmic-ray irradiation processes. The roles played by the above-mentioned three processes are beautifully displayed and they are represented by three different groups of meteorites, as shown in Tables 7.6, 7.7 and 7.8. When Sir William Ramsay discovered a new rare gas element in June 1898, he named it krypton or "hidden". It seems as if the key to the solution of the puzzle of the krypton had been carefully hidden for so many years since the 1960's.

7.7. Xenon

Xenon has nine stable isotopes: ^{124}Xe, ^{126}Xe, ^{128}Xe, ^{129}Xe, ^{130}Xe, ^{131}Xe, ^{132}Xe, ^{134}Xe and ^{136}Xe. The heaviest isotope ^{136}Xe is a magic nuclide containing 82 neutrons. As it has been discussed in Chapter 6, Section 6.9 (Unsolved Problems in Xenology), the case of xenon is extremely complex not only because it has many stable isotopes, but also because its isotopic compositions are often altered by the addition of the decay products of ^{129}I and ^{244}Pu, which existed in nature during the early history of the solar system (see Chapter 6).

The relationship between the isotopic compositions of the meteoritic and terrestrial (atmospheric) xenon can be written in the form

$$(1 + f_{136} + n_{136} + c_{136}) \cdot M^{124} = A^{124} \cdot (1 + 12\mu) + n_{124} + c_{124}$$
$$(1 + f_{136} + n_{136} + c_{136}) \cdot M^{126} = A^{126} \cdot (1 + 10\mu) + n_{126} + c_{126}$$
$$(1 + f_{136} + n_{136} + c_{136}) \cdot M^{128} = A^{128} \cdot (1 + 8\mu) + n_{128} + c_{128}$$
$$(1 + f_{136} + n_{136} + c_{136}) \cdot M^{129} = A^{129} \cdot (1 + 7\mu) + n_{129} + c_{129} + r_{129} \qquad (7.14)$$
$$(1 + f_{136} + n_{136} + c_{136}) \cdot M^{130} = A^{130} \cdot (1 + 6\mu) + n_{130} + c_{130}$$
$$(1 + f_{136} + n_{136} + c_{136}) \cdot M^{131} = A^{131} \cdot (1 + 5\mu) + n_{131} + c_{131} + f_{131}$$
$$(1 + f_{136} + n_{136} + c_{136}) \cdot M^{132} = A^{132} \cdot (1 + 4\mu) + n_{132} + c_{132} + f_{132}$$
$$(1 + f_{136} + n_{136} + c_{136}) \cdot M^{134} = A^{134} \cdot (1 + 2\mu) + n_{134} + c_{134} + f_{134}$$

where M^i and A^i are the $^iXe/^{136}Xe$ ratios in the meteorite and in the earth's atmosphere, respectively: μ is a mass-fractionation factor; n_i is the contribution from neutron-capture processes at mass number i; c_i is the contribution from the cosmic-ray-produced isotope at mass number i; r_{129} is the contribution from the decay of ^{129}I; and the f_i's are the contributions from the spontaneous fission decay products of ^{244}Pu at mass numbers $i = 131$, 132, 134 and 136.

Both the ratios of $f_{131} : f_{132} : f_{134} : f_{136}$ and $c_{124} : c_{126} : c_{128} : c_{129} : c_{130} : c_{131} : c_{132}$ are known (see Chapter 6, Section 6.8 and Table 6.8). The value of c_{134} is about one tenth of c_{132} and hence can be neglected in most cases. The value of c_{136} can be considered as zero and the values of n_{134} and n_{136} are also negligibly small, because the neutron-capture cross-sections of ^{134}Xe and ^{136}Xe are both very small.

In the cases of meteorites with high gas contents (carbonaceous and gas-rich chondrites), the relationship given by (7.14) therefore can be written in a simplified form

$$M^{124} = A^{124} \cdot (1 + 12\mu) + n_{124} + c_{124}$$
$$M^{126} = A^{126} \cdot (1 + 10\mu) + n_{126} + c_{126}$$
$$M^{128} = A^{128} \cdot (1 + 8\mu) + n_{128} + c_{128}$$
$$M^{129} = A^{129} \cdot (1 + 7\mu) + n_{129} + c_{129} + r_{129}$$
$$M^{130} = A^{130} \cdot (1 + 6\mu) + n_{130} + c_{130} \qquad (7.15)$$
$$M^{131} = A^{131} \cdot (1 + 5\mu) + n_{131} + c_{131}$$
$$M^{132} = A^{132} \cdot (1 + 4\mu) + n_{132} + c_{132}$$
$$M^{134} = A^{134} \cdot (1 + 2\mu)$$

Table 7.9 shows the isotopic compositions of xenon in carbonaceous chondrites and ureilites [7.55, 7.56, 7.61–7.64]. The xenon isotope data are expressed here in terms of

$$\delta i = \left\{ (^iXe/^{130}Xe)_M / (^iXe/^{130}Xe)_A - 1 \right\} \cdot (1{,}000) \text{ (per mil)} \qquad (7.16)$$

7.61   O.K. Manuel, R.J. Wright, D.K. Miller and P.K. Kuroda, J. Geophys. Res. 75:5693 (1970)
7.62   L.L. Wilkening and K. Marti, Geochim. Cosmochim. Acta 40:1465 (1976)
7.63   P.K. Kuroda, R.D. Sherrill, D.W. Efurd, and J.N. Beck, J. Geophys. Res. 80:1558 (1975)
7.64   P.K. Kuroda, J.N. Beck, D.W. Efurd and D.K. Miller, J. Geophys. Res. 79:3981 (1974)

Table 7.9. Isotopic composition of xenon in carbonaceous chondrites and ureilites

Meteorite	δ124	δ126	δ128	δ129	δ130	δ131	δ132	δ134	δ136	^{130}Xe content $(10^{-12}$ cc STP/g)
(1) Allende	+207 ± 30	+174 ± 23	+114 ± 13	+1088 ± 28	≡ 0	−31.6 ± 9.6	−62.8 ± 9.4	−51.2 ±10.6	− 39.6 ± 6.9	268
(2) Mokoia	+228 ± 21	+206 ± 23	+ 78.0 ± 12.8	+ 257 ± 13	≡ 0	−31.6 ± 9.6	−69.0 ± 8.3	−62.1 ±10.2	− 51.6 ± 7.4	436
(3) Leoville	+139 ± 21	+102 ± 23	+ 65.3 ± 12.8	+ 24.7 ± 9.2	≡ 0	−29.1 ±11.6	−63.9 ± 6.1	−77.8 ±11.7	− 80.0 ± 9.2	697
(4) Kenna	+227 ± 13	+152 ± 9	+ 80.2 ± 12.8	− 18.8 ± 2.0	≡ 0	−31.6 ± 1.9	−69.0 ± 1.8	−96.1 ± 2.0	−119 ± 3	1063
(5) Murchison	+241 ± 17	+165 ± 9	+ 76.5 ± 7.4	+ 4.5 ± 5.1	≡ 0	−28.8 ± 4.8	−66.9 ± 3.8	−83.2 ± 4.7	− 93.3 ± 5.5	2015
(6) Murray	+267 ± 21	+179 ± 23	+ 82.3 ± 14.9	− 18.2 ± 12.7	≡ 0	−16.8 ±10.8	−64.6 ±10.3	−80.9 ±12.1	− 91.5 ± 12.0	2383
(7) Novo Urei	+193 ± 40	+149 ± 27	+ 64.4 ± 12.8	− 21.6 ± 7.7	≡ 0	−27.2 ± 7.7	−59.9 ± 5.6	−77.4 ± 9.4	− 87.9 ± 10.1	3712
(8) Neutron-irradiated Atmospheric Xe: Φt = 1 × 10^{22} (n/cm²) (10 KeV)[a]	− 74.9	− 74.7	+ 90.8	− 73.3	≡ 0	−74.2	−58.9	−66.2	− 65.2	—

[a] A mixture of Xe, I, and Te in the ratio of their cosmic abundances given by Burbidge et al. [7.5] was assumed to have been exposed to 10 KeV neutrons. The neutron fluence was assumed to have been such that the abundance of ^{136}Xe became depleted by 0.406 parts per 10^4

135

where the subscripts M and A refer to the meteoritic and atmospheric xenon. Note that they resemble the isotopic composition of a neutron-irradiated atmospheric xenon. The isotopic composition of the neutron-irradiated atmospheric xenon was calculated assuming that a mixture of xenon, iodine, and tellurium in the ratio of their cosmic abundance values was exposed to 10 KeV neutrons. The neutron fluence was assumed to have been such that the abundance of ^{136}Xe became depleted by 0.406 parts per 10^4. The 10 KeV cross-section values reported by Holmes et al. [7.65] were used in the calculations. The 10 KeV cross-section value for ^{136}Xe is 0.00406 barns and hence the value of 10 KeV neutron fluence assumed here corresponds to 1×10^{22} (n/cm^2).

These results indicate that the meteorites contain xenon, which had been subjected to a much more intense neutron irradiation than the terrestrial xenon. It is quite possible that we are dealing here with the so-called solar xenon, which was transported from the sun in the form of solar wind and was incorporated into these meteorites. The solar xenon may have been subjected to a high neutron flux during the deuterium-burning stage of the sun shortly after its birth, as was suggested by Cameron [7.66] in 1964.

It is also worthy of note in Table 7.9 that there is a negative correlation between the concentration of xenon and the value of $\delta 129$. The existence of such a relationship may be regarded as an indication that xenon was being removed from solid bodies during and after an early irradiation period.

The 10 KeV neutron-capture cross-section of ^{128}Te is 197 mb and the cosmic abundance values of ^{128}Te and ^{127}I are 1.48 and 0.80, respectively. If it is assumed that the extinct nuclide ^{129}I was produced by neutron-capture reactions on ^{128}Te in a mixture of iodine and tellurium in the ratio of their cosmic abundance values and if it is assumed that the neutron fluence was 1×10^{22} (n/cm^2), a simple calculation yields a value of

$$\frac{^{129}I}{^{127}I}_o = \frac{(197)(10^{-24})(10^{22})(1.48)}{(0.80)}$$
$$= 3.6 \times 10^{-3} \text{ (atom/atom)}$$

for the initial ratio of $^{129}I/^{127}I$.

The value of the initial $^{129}I/^{127}I$ ratio thus calculated is considerably greater than the corresponding ratio for the ^{129}I, which is assumed to have originated in the r-processes occurring in supernovae. For example, Kuroda [7.67] reported in 1961 a value of 1×10^{-3} (atom/atom) for the initial ratio of $^{129}I/^{127}I$ at the time of the cessation of nucleosynthesis in stars.

Results from these calculations show that ^{129}I was probably produced by neutron-capture reactions on ^{128}Te during an early irradiation period, as was originally suggested by Fowler et al. [7.6] in 1962:

$$^{128}\text{Te}(n,\gamma)^{129}\text{Te} \xrightarrow[\text{70m; 33.4d}]{\beta^-} {}^{129}\text{I} \xrightarrow[\text{1.59} \times 10^7 \text{ y}]{\beta^-} {}^{129}\text{Xe (stable)}$$

7.65   J.A. Holmes, S.E. Woosley, W.A. Fowler and B.A. Zimmerman, Atomic Data and Nuclear Data Tables 18:305 (1976)
7.66   A.G.W. Cameron, The Origin and Evolution of Atmospheres and Oceans, edited by P.J. Brancazio and A.G.W. Cameron, John Wiley & Sons, Inc., New York, 1964, p. 235
7.67   P.K. Kuroda, Geochim. Cosmochim. Acta 24:40 (1961)

In 1960, while attempting to explain the differences in the isotopic compositions of the rare gases found in the meteorites Richardton, Murray and in the earth's atmosphere, Reynolds [7.68] noted that a strong mass-dependent fractionation may have been responsible for most of the xenon isotope anomalies. According to this interpretation, however, it appeared as if there was excess ^{134}Xe and ^{136}Xe in meteoritic xenon, which could be attributed to the presence of a fission xenon component (see Chapter 6, Section 6.5). This idea was pursued by several investigators and led to the general conclusion that the carbonaceous and gas-rich chondrites contain large excesses of an unexplained fission component [7.69—7.74], which in turn resulted in the interesting hypothesis that this fission xenon component may be the spontaneous fission decay product of an unknown superheavy nuclide [7.75—7.80].

Fig. 7.2 shows the difference in the isotopic compositions of xenon in the carbonaceous chondrite Murray and in the earth's atmosphere. The δi values for i = 124, 126, 128 und 130 lie along the straightline mm', which gives the isotopic composition of a severely mass-fractionated atmospheric xenon. The values of δi for i = 131, 132, 134 and 136 lie above the line mm', however, and it appears as if there is an excess of fission xenon in the meteorite.

According to this interpretation, the relationship between the isotopic compositions of the meteoritic and terrestrial (atmospheric) xenon should be written in the form

$$
\begin{aligned}
(1+f_{136}) \cdot M^{124} &= A^{124} \left(1 + 12\mu\right) \\
(1+f_{136}) \cdot M^{126} &= A^{126} \left(1 + 10\mu\right) \\
(1+f_{136}) \cdot M^{128} &= A^{128} \left(1 + 8\mu\right) \\
(1+f_{136}) \cdot M^{129} &= A^{129} \left(1 + 7\mu\right) + r_{129} \\
(1+f_{136}) \cdot M^{130} &= A^{130} \left(1 + 6\mu\right) \\
(1+f_{136}) \cdot M^{131} &= A^{131} \left(1 + 5\mu\right) + f_{131} \\
(1+f_{136}) \cdot M^{132} &= A^{132} \left(1 + 4\mu\right) + f_{132} \\
(1+f_{136}) \cdot M^{134} &= A^{134} \left(1 + 2\mu\right) + f_{134}
\end{aligned}
\tag{7.17}
$$

The equation (7.17) predicts that the values of δi for i = 124, 126, 128 should fall along a straight line, if the data are plotted in the manner shown in Fig. 7.2. It turns out, however, that the experimental data fail to display such a straightline relationship.

7.68   J.H. Reynolds, Phys. Rev. Lett. 4:8, 351 (1960)
7.69   R.O. Pepin, The Origin and Evolution of Atmospheres and Oceans, edited by P.J. Brancazio and A.G.W. Cameron, John Wiley & Sons, Inc., New York, 1964, p. 191
7.70   R.O. Pepin, Origin and Distribution of the Elements, edited by L.H. Ahrens, Pergamon Press, 1968, p. 379
7.71   H. Funk, F. Podosek and M.W. Rowe, Geochim. Cosmochim. Acta 31:1721 (1967)
7.72   K. Marti, Science 166:1263 (1969)
7.73   M.W. Rowe, Geochim. Cosmochim. Acta 32:1317 (1968)
7.74   N. Takaoka, Mass Spectrom. 20:287 (1972)
7.75   E. Anders and D. Heymann, Science 164:821 (1969)
7.76   M. Dakowski, Earth Planet. Sci. Letters 6:152 (1969)
7.77   B. Srinivasan, E.C. Alexander, Jr., O.K. Manuel and D.E. Troutner, Phys. Rev. 179:1166 (1969)
7.78   M.N. Rao, Nuclear Physics A 140:69 (1970)
7.79   E. Mazor, D. Heymann and E. Anders, Geochim. Cosmochim. Acta 34:781 (1970)
7.80   E. Anders and J.W. Larimer, Science 175:981 (1972)

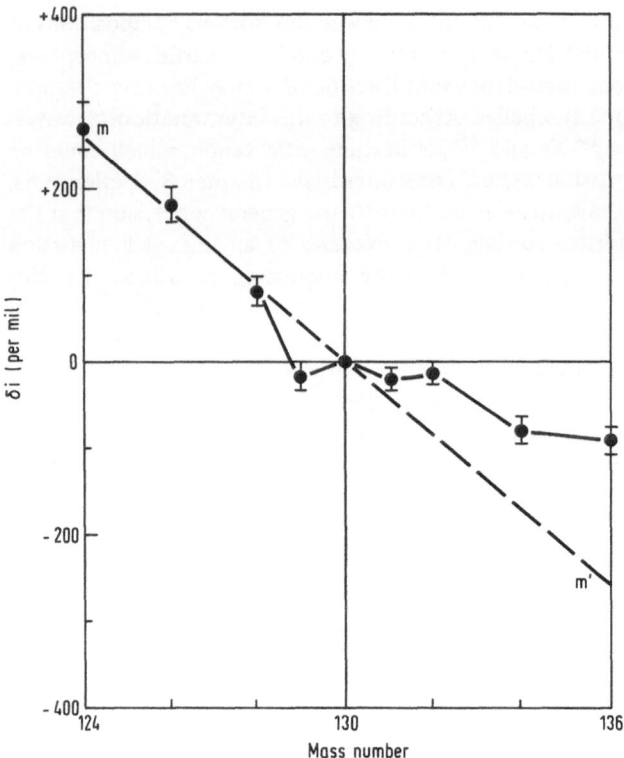

Fig. 7.2. The Puzzle of the Xenon: I. The difference in the isotopic compositions of xenon in the carbonaceous chondrite Murray and in the earth's atmosphere is shown here. If it is assumed that a strong mass-dependent fractionation (shown by the line mm') was responsible for most of the anomalies, the data suggests that there is an excess of fission xenon in Murray: the so-called carbonaceous chondrite fission (CCF) xenon.

In 1975, Lewis et al. [7.60] reported that they found the host phase of the xenon in the carbonaceous chondrite Allende. They subjected the meteorite to acid-etching experiments and found the acid-insoluble residues to be highly enriched in CCF. Fig. 7.3 shows the isotopic composition of the xenon which they found in the Allende residue 3CS4. Note that although the relative abundances of ^{134}Xe and ^{136}Xe are sharply enhanced in Allende 3CS4, the abundances of the light isotopes ^{124}Xe, ^{126}Xe and ^{128}Xe are also greatly enhanced. Moreover, the value of δi for ^{132}Xe is negative, indicating that there may be an excess of ^{130}Xe in the residue. Lewis et al. [7.60] stated that the enhancement of the light isotopes could be attributed to some kind of mass-fractionation process and they deduced a fission spectrum, in which the yields at mass numbers 131 and 132 were abnormally small. Manuel et al. [7.32] argued, on the other hand, that these xenon fractions may contain material that has been added to our solar system from a nearby supernova (see Section 7.2). It appears, however, that these wild speculations lead us nowhere.

It seems that these strange xenon fractions were actually "manufactured" by subjecting the meteorite samples to stepwise-heating and acid-etching procedures. Cherdyntsev

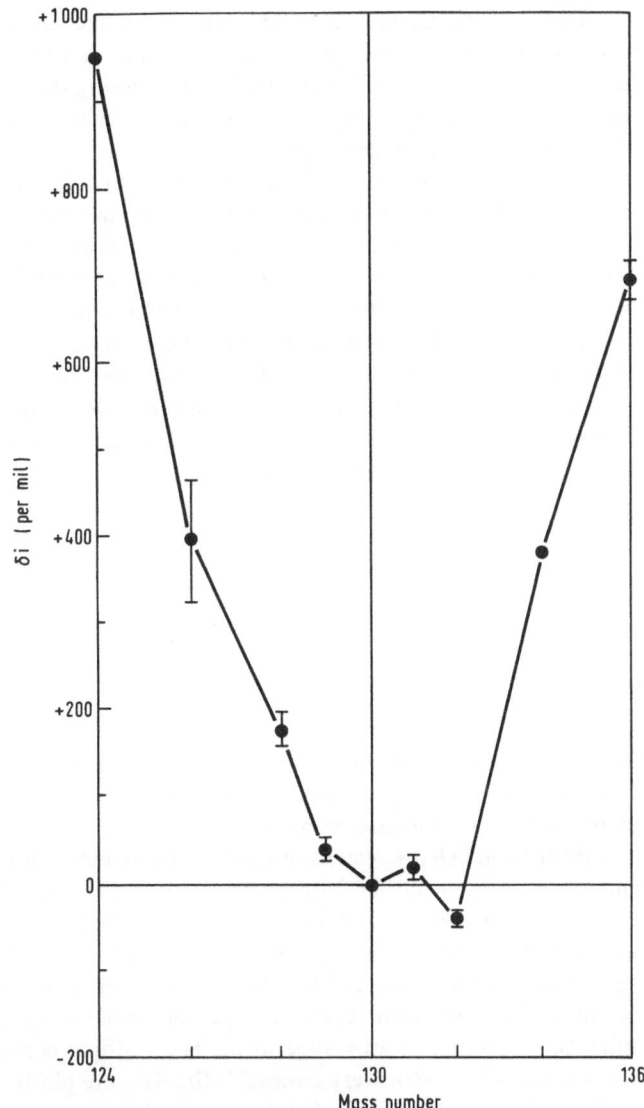

Fig. 7.3. The Puzzle of the Xenon: II. The isotopic composition of the xenon found in the Allende residue 3CS4 by Lewis et al. [7.60] is shown here. It appears as if the CCF xenon is highly enriched in this residue, but the enhancement of the light isotopes can not be explained

[7.81] reported that the ^{234}U/^{238}U ratio does not remain constant in an extract from a mineral. They found, for example, uranium in a zircon extract showed the ^{234}U/^{238}U = 1.8 (curie/curie) and for uranium of surface waters, the observed ratio of ^{234}U/^{238}U equalled 7 to 8 (curie/curie). Uranium sublimed from pitchblende at 800 °C was enriched in ^{234}U to the extent of ^{234}U/^{238}U = 3.2 (curie/curie).

7.81   V.V. Cherdyntsev, Abundance of Chemical Elements, translated by W. Nichiporuk, The University of Chicago Press, 1961, p. 107

These results can be explained as due to the fact that the decayed atoms (^{234}U), after recoil, pass into different sites of the crystalline lattice, where they reside in the form of foreign, generally weakly bound, inclusions. During the interaction of natural waters with minerals, the ^{234}U atoms pass into solution much more easily than do the isotopes, which did not undergo similar decay (^{238}U).

In the case of the xenon isotopes in meteorites, it is quite probable that the xenon atoms created by different nuclear processes are not bound to the different mineral phases with equal force. When a meteorite sample is subjected to stepwise-heating or acid-etching experiments, these different xenon isotopes are expected to be released in a very complex manner and hence one must be extremely careful in interpreting the experimental data.

The isotopic composition (R^i) of the strange xenon found in the Allende residue 3CS4 may be explained in the following manner: Suppose that the isotopic composition of xenon which was originally present in a sample of meteorite was given by Q^i and that of a part of the original xenon which was lost during the acid treatment was represented by P^i. Then we have the relationship

$$Q^i = (1 - \psi) \cdot P^i + \psi \cdot R^i$$

or

$$R^i = \frac{1}{\psi} \left\{ Q^i - (1 - \psi) \cdot P^i \right\} \tag{7.18}$$

where P^i, Q^i and R^i refer to the $^iXe/^{136}Xe$ ratios; ψ is the ratio of the number of atoms of ^{136}Xe found in the residue to that of ^{136}Xe, which was originally present in the untreated meteorite; and i is the mass number.

Both P^i and Q^i are related to the isotopic composition (A^i) of the atmospheric xenon and can be written in the form shown by equations (7.15). The isotopic composition of the xenon in the residue (R^i) is also related to the isotopic composition (A^i) of the atmospheric xenon and can be explained in terms of μ, the n^i's and the c_i's. In other words, the xenon found in the residue contains neither an exotic substance such as the decay product of an unknown superheavy element nor the material which was injected to the solar nebula from a nearby supernova. The isotopic composition of the xenon found in the residue appears to be very unusual if the data are plotted in the manner which is shown in Fig. 7.3, when the value of ψ is very small compared to 1. For example, the Allende meteorite studied by Lewis et al. [7.60] contained 743×10^{-12} cc STP of ^{136}Xe per gram. The amount of ^{136}Xe found in the residue 3CS4 was 19×10^{-12} cc STP and hence $\psi = 0.0295$.

7.8. Barium

Barium has seven stable isotopes: ^{130}Ba, ^{132}Ba, ^{134}Ba, ^{135}Ba, ^{136}Ba, ^{137}Ba and ^{138}Ba. The heaviest isotope ^{138}Ba is a magic nuclide containing 82 neutrons. The case of barium with $Z = 56$ is interesting because it is the next even Z element after xenon with $Z = 54$. The relationship between the isotopic compositions of meteoritic and terrestrial (normal) barium can be written in the form

$$(1 + n_{138}) \cdot M^{130} = T^{130} (1 + 8\mu) + n_{130}$$
$$(1 + n_{138}) \cdot M^{132} = T^{132} (1 + 6\mu) + n_{132}$$
$$(1 + n_{138}) \cdot M^{134} = T^{134} (1 + 4\mu) + n_{134}$$
$$(1 + n_{138}) \cdot M^{135} = T^{135} (1 + 3\mu) + n_{135} \qquad (7.19)$$
$$(1 + n_{138}) \cdot M^{136} = T^{136} (1 + 2\mu) + n_{136}$$
$$(1 + n_{138}) \cdot M^{137} = T^{137} (1 + 1\mu) + n_{137}$$

where M^i and T^i refer to the $^iBa/^{138}Ba$ ratios in meteoritic and terrestrial (normal) samples, respectively; μ is a mass-fractionation factor; and n_i is the contribution from the neutron-capture processes at mass number i.

In 1962, Umemoto [7.82] measured the isotopic compositions of barium extracted from terrestrial samples and from the meteorites Bruderheim, Pasamonte and Nuevo Laredo. He reported that the meteorites contained more of the low mass isotopes relative to the highest mass isotope of barium than did the terrestrial samples and the variation was approximately linear with mass. Murthy [7.83] pointed out in 1964 that the observation by Umemoto, if real, was very important because it was the first indication that the so-called general anomalies existed in meteorites for the non-volatile heavy elements as well.

Table 7.10. A comparison of the isotopic compositions of barium in meteorites and of a neutron-irradiated normal barium

Sample	$\epsilon130$	$\epsilon132$	$\epsilon134$	$\epsilon135$	$\epsilon136$	$\epsilon137$	$\epsilon138$
(1) Allende EK1-4-1SCa[a]	− 1.2 ± 7.5	− 2.5 ±14.2	≡ 0	+13.8 ± 0.9	≡ 0	+13.5 ± 0.4	+ 1.6 ± 0.9
(2) Bruderheim[b]	+31.2 ±13.6	+13.7 ±20.4	≡ 0	−13.8 ± 8.6.	≡ 0	−21.6 ± 8.9	− 7.7 ± 9.9
(3) Normal Ba exposed to 10 KeV neutrons: $\Phi t = 2.1 \times 10^{21}$ (n/cm²)[c]	+31.2	+16.3	≡ 0	−18.9	≡ 0	−22.4	−30.7

[a] Calculated from the experimental data reported by McCulloch and Wasserburg (1978) [7.25]
[b] Umemoto (1962) [7.82]
[c] The 10 KeV cross-section values reported by Holmes et al. [7.65] were used in these calculations. The neutron fluence was assumed to be such that ^{138}Ba became depleted by 0.20 parts per 10^4

Table 7.10 compares the barium isotope data obtained by Umemoto [7.82] in 1962 and by McCulloch and Wasserburg [7.25] in 1978. The latter authors have normalized the barium isotope data in such a manner that for isotope i as deviations from normal ratios in parts in 10^4 with ^{138}Ba as the index isotope:

$$\epsilon i = [(^iBa/^{138}Ba)_C / (^iBa/^{138}Ba)_N - 1] \cdot 10^4 \qquad (7.20)$$

7.82   S. Umemoto, J. Geophys. Res. 67:375 (1962)
7.83   V.R. Murthy, Isotopic and Cosmochemistry, edited by H. Craig, S.L. Miller and G.J. Wasserburg, North-Holland Pub. Co., Amsterdam, 1964, p. 488

where the subscript N refers to the normal barium and the subscript C to the ratios measured and normalized for mass fractionation assuming ^{135}Ba/^{138}Ba to equal the terrestrial value. In Table 7.10, the data are re-normalized to the shielded isotopes ^{134}Ba and ^{136}Ba in such a manner that ϵ 134 = 0 and ϵ 136 = 0.

Note that the isotopic composition of barium from the Bruderheim meteorite is quite similar to that of normal barium irradiated with neutrons. The 10 KeV cross-section values reported by Holmes et al. [7.65] were used here and the neutron fluence was assumed to have been such that ^{138}Ba became depleted by 0.20 parts per 10^4. In calculating the values of ϵi for the normal barium exposed to a neutron flux, the contributions from the neutron-capture reactions on ^{133}Cs, ^{134}Xe, ^{136}Xe, and ^{136}Ce were assumed to have been negligibly small.

Table 7.10 shows that the isotopic composition of barium extracted from the Bruderheim meteorite is quite similar to that of normal barium irradiated with 10 KeV neutrons, while the ϵi values for the barium extracted from the Allende inclusion and the Bruderheim meteorite are opposite in sign. These results can be interpreted as being due to the fact that the Allende inclusions were exposed to a smaller fluence of 10 KeV neutrons than the average earth material, while the barium in the Bruderheim meteorite was exposed to a larger fluence of 10 KeV neutrons than the average earth material. The 10 KeV neutron-capture cross-section of ^{138}Ba is 9.46 mb and hence the difference in the neutron fluence is calculated to be roughly equal to 2.1×10^{21} (n/cm^2).

7.9. Gadolinium

Gadolinium has seven stable isotopes: ^{152}Gd, ^{154}Gd, ^{155}Gd, ^{156}Gd, ^{157}Gd, ^{158}Gd and ^{160}Gd. The lightest isotope ^{152}Gd is actually radioactive and its alpha half-life is 1.1×10^{14} years. The case of gadolinium is interesting because the thermal neutron-capture cross sections of ^{155}Gd and ^{157}Gd are very large. As in the case of barium, the relationship between the isotopic compositions of meteoritic and terrestrial (normal) gadolinium can be written in the form

$$
\begin{aligned}
(1 + n_{160}) \cdot M^{152} &= T^{152} (1 + 8\mu) + n_{152} \\
(1 + n_{160}) \cdot M^{154} &= T^{154} (1 + 6\mu) + n_{154} \\
(1 + n_{160}) \cdot M^{155} &= T^{155} (1 + 5\mu) + n_{155} \\
(1 + n_{160}) \cdot M^{156} &= T^{156} (1 + 4\mu) + n_{156} \\
(1 + n_{160}) \cdot M^{157} &= T^{157} (1 + 3\mu) + n_{157} \\
(1 + n_{160}) \cdot M^{158} &= T^{158} (1 + 2\mu) + n_{158}
\end{aligned}
\tag{7.21}
$$

where M^i and T^i refer to the ^{i}Gd/^{160}Gd ratios in meteoritic and terrestrial (normal) samples, respectively; μ is a mass-fractionation factor; and n_i is the contribution from the neutron-capture processes at mass number i.

As it was mentioned in Section 7.1., Eugster et al. [7.8], in 1970, found neutron-capture effects in gadolinium extracted from the Norton County achondrite, but they concluded that these effects were most probably produced by secondary neutrons during a long cosmic-ray exposure of this large meteorite. They added that the observed isotopic anomalies corresponded to an integrated thermal neutron flux of $(6.3 \pm 0.9) \times 10^{15}$ neutrons per cm^2.

Table 7.11. Relative abundances (N) of the isotopes of Gd and their neutron-capture cross-sections

	(^{151}Eu)	^{152}Gd	(^{153}Eu)	^{154}Gd	(^{154}Sm)
N	(0.5785)	0.00928	(0.6550)	0.09975	(1.0067)
σ (thermal)	9 × 10^3 [a]	10	3.8 × 10^2	90	5
σ (15 KeV)	5.6[b]	2.4	5.6	2.4	1.44

	^{155}Gd	^{156}Gd	^{157}Gd	^{158}Gd	^{160}Gd
N	0.67692	0.9361	0.71589	1.13590	1.00000
σ (thermal)	6.1 × 10^4	2	2.55 × 10^5	2.4	0.77
σ (15 KeV)	6.8	2.4	6.8	2.4	2.4

[a] Neutron capture reactions on ^{151}Eu lead to 9.3-h ^{152}Eu and 12-y ^{152}Eu. The former decays 77% of the time and the latter 28% of the time to ^{152}Gd. The cross-section values for the production of 9.3-h ^{152}Eu and 12-y ^{152}Eu are 3.2 × 10^3 and 5.8 × 10^3 barns, respectively
[b] The neutron-capture cross-section value leading to the formation of ^{152}Gd was calculated from the total cross-section value assuming that the isomeric production ratio is the same as that for thermal neutrons

In 1979, Kuroda [7.84] pointed out, however, that neutron-capture effects observed by Eugster et al. [7.8] could be explained alternatively as due to the fact that the gadolinium in the Norton County meteorite was exposed to a greater non-thermal neutron flux than the average earth material. Table 7.11 shows the relative abundances of the gadolinium isotopes and their neutron-capture cross-section values. The values for ^{151}Eu, ^{153}Eu, and ^{154}Sm are also included here, since these isotopes are expected to contribute to the alteration of the abundances of ^{152}Gd, ^{154}Gd, and ^{155}Gd, respectively. The cosmic abundance values reported by Burbidge et al. [7.5] were used here to calculate the abundances of the Eu and Sm isotopes relative to ^{160}Gd, whose abundance was taken to be equal to 1.00000.

Table 7.12 compares the isotopic composition of the Gd extracted from the Norton County achondrite with those for the normal Gd exposed to thermal and 15 KeV neutrons. The isotopic composition is expressed here in terms of ϵi defined as

$$\epsilon i = \left\{ (^iGd/^{160}Gd)_C/(^iGd/^{160}Gd)_N - 1 \right\} \cdot 10^4 \tag{7.22}$$

where the subscript N refers to the normal gadolinium and the subscript C to the ratios measured and normalized for mass-fractionation, assuming the variation in the ^{156}Gd/^{160}Gd ratio to be entirely due to mass-fractionation.

Table 7.12 shows that the observed anomalies agree reasonably well with the calculated values for the neutron-irradiated gadolinium using a value of $\Phi t = 6.3 \times 10^{15}$ (n/cm^2), but the agreement is poor at mass number 152. It is worthy of note here that Eugster et al. [7.8] stated that a samarium contamination was present in these samples and, while

7.84   P.K. Kuroda, Geochem. J. 13:281 (1979)

Table 7.12. Isotopic composition of Gd in the Norton County achondrite

Sample	ε 152	ε154	ε155	ε156	ε157	ε158	ε160
(1) Norton County A[a]	+ 53.9	− 4.0	− 9.0	≡ 0	−18.9	+ 8.5	≡ 0
	± 10.8	± 4.0	± 4.4		± 4.7	± 2.2	
(2) Norton County B[a]	+172.4	−10.0	− 4.9	≡ 0	−18.4	+ 8.8	≡ 0
	± 43.1	±15.0	± 6.4		± 3.9	± 5.0	
(3) Neutron-irradiated normal Gd							
a) $\Phi t = 6.3 \times 10^{15}$ (n/cm²) (thermal)	+ 14.7	− 5.6	−10.3	≡ 0	−25.3	+12.2	≡ 0
b) $\Phi t = 1 \times 10^{20}$ (n/cm²) (15 KeV)	+150	+29.0	− 8.6	≡ 0	− 4.9	+ 1.8	≡ 0

[a] Calculated from the experimental data reported by Eugster et al. [7.8] in 1970

the ^{152}Gd/^{160}Gd ratios shown in Table 7.12 were uncorrected for any possible interferences, the ^{154}Gd/^{160}Gd ratios were always corrected for ^{154}Sm. They further noted that the samarium correction to ^{154}Gd was always less than 0.3 percent.

In comparing the gadolinium anomalies observed in the Norton County achondrite with normal gadolinium exposed to neutrons, it is therefore important to keep in mind the fact that the ^{154}Gd/^{160}Gd ratios shown in Table 7.12 are the corrected values. If the samarium correction to ^{154}Gd was as much as 0.3 percent or 30 parts per 10^4, the observed values of ε154 for the Norton County achondrites A and B must have been +26.0 ± 4.0 and +20.0 ± 15.0, instead of −4.0 ± 4.0 and −10.0 ± 15.0, respectively, and if so, it seems to be more reasonable to conclude that the neutron effects observed in the Norton County achondrite were produced by a total flux of about 1×10^{20} (n/cm²) of 15 KeV neutrons, rather than by 6.3×10^{15} (n/cm²) of thermal neutrons.

7.10. Other Elements

McCulloch and Wasserburg [7.25] and Lugmair [7.26] reported on the isotopic anomalies in neodymium extracted from the Allende inclusions. These anomalies are similar in nature to those observed in samarium and gadolinium and can be explained as due to the fact that the Allende inclusions have received a smaller neutron fluence than the average earth material during an early irradiation period [7.85].

Umemoto [7.82] found that the isotopic composition of cerium extracted from the Bruderheim meteorite was different from the normal terrestrial cerium. The anomalies observed in cerium seem to be similar to the case of barium and the data indicate that the Bruderheim meteorite was subjected to a greater total neutron flux than the average earth material during an early irradiation period [7.86].

7.85   P.K. Kuroda, Geochem. J. 13:291 (1979)
7.86   P.K. Kuroda, Geochem. J. 13:137 (1979)

The isotopic anomalies observed in calcium and titanium in the Allende inclusions appear to be related to each other and can be explained as due to a combined effect of mass-fractionation and cosmic-ray irradiation processes which occurred during the early history of the solar system [7.87].

Kelly and Wasserburg [7.28] found the ^{107}Ag/^{109}Ag ratio in the Santa Clara iron meteorite to be 4 percent greater than the terrestrial value. They attributed the enrichment of ^{107}Ag to the decay of ^{107}Pd:

$$^{107}\text{Pd} \xrightarrow[6.5 \times 10^6 \text{ year}]{\beta^-} {}^{107}\text{Ag (stable)}$$

which occurred during the early history of the solar system.

As was pointed out by Kuroda [7.88], it appears that ^{107}Pd was produced by the ^{106}Pd(n,γ)^{107}Pd reactions during an early irradiation period. Suppose that the palladium isotopes in the iron meteorite were irradiated with a fluence of 1×10^{20} (n/cm^2) of 15 KeV neutrons. The 15 KeV neutron-capture cross-section of ^{106}Pd is 0.504 barns and hence the fraction of ^{106}Pd which will be converted to ^{107}Pd by the neutron-capture reactions is

$$(1 \times 10^{20})(0.504 \times 10^{-24}) = 5.04 \times 10^{-5}$$

or 5.04 parts per 10^5.

This value is in agreement with the initial ratio of ^{107}Pd/^{110}Pd equal to or greater than 2×10^{-5} (atom/atom), which was reported by Kelly and Wasserburg [7.28]. These values are also in agreement with the results of their measurements on the isotopic composition of palladium in the Santa Clara meteorite. The ^{106}Pd/^{104}Pd ratio in the meteorite agreed with that for normal palladium within about $\pm$ 20 parts per 10^5.

It is interesting to note here that the unified theory of isotopic anomalies in meteorites enables one to predict the nature and the magnitude of the variations in the isotopic compositions expected to be observed for the remaining chemical elements in the Periodic Table. The case of uranium anomalies is of special interest. Arden [7.35] has reported that measurements of the isotopic composition of uranium in five meteorites (Parnallee, Barwell, Allende, Bruderheim, and Richardton) showed that variations in the uranium isotopic composition existed in these chondrites, indicating the presence of components of low ^{238}U/^{235}U ratio. Tatsumoto and Shimamura [7.89] have also found large variations in ^{235}U/^{238}U of from +40 to −35 permil in the inclusions of the Allende meteorite and variations of from +434 to −157 permil in the HCl-HF residue of the Allende matrix. They interpreted these variations as strong evidence for *live* ^{247}Cm in the early solar system. Chen and Wasserburg [7.90] have recently reported, however, that the total range in ^{235}U/^{238}U in the acid-soluble phases of the Allende inclusions and meteoritic whitlockites was from 1/137.6 to 1/138.3 and to within the error limits (3 to 20 permil) indistinguishable from normal terrestrial uranium.

7.87 P.K. Kuroda, Geochem. J. 13:297 (1979)
7.88 P.K. Kuroda, Geochem. J. 13:135, 291 (1979)
7.89 M. Tatsumoto and T. Shimamura, Nature 286:118 (1980)
7.90 J.H. Chen and G.J. Wasserburg, Earth Planet. Sci. Lett. 52:1 (1981)

It is worthy of note here that large variations in the uranium isotopic ratios were observed only when the samples were subjected to repeated acid-etching experiments, just as in the case of xenon isotopes whereby xenon fractions with very strange isotopic compositions were isolated from meteorites by subjecting the samples to stepwise-heating experiments, repeated acid-etching experiments or a combination of both (see Section 7.7., Xenon).

As was mentioned earlier, the anomalies observed in the samarium from the Allende inclusions indicate that the inclusions have received a smaller integrated flux of nonthermal neutrons than the average solar system material and the difference was equivalent to a fluence of 3.3×10^{20} (n/cm^2) of 15 KeV neutrons (see Table 7.4 in Section 7.3). The cross section values for the (n,f) and n,γ) reaction on ^{235}U at 15 KeV are about 3 and 1 barns, respectively, while the corresponding values for ^{238}U are 0 and about 0.5 barns, respectively (see, for example, G.R. Choppin and J. Rydberg, Nuclear Chemistry: Theory and Applications, Pergamon Press (1980), p. 446). It is therefore to be expected that the abundance of ^{235}U in the Allende inclusions will be enhanced relative to that in the average solar system material by

$$(4 \times 10^{-24})(3.3 \times 10^{20}) - (3.1)(0.5 \times 10^{-24})(3.3 \times 10^{20}) =$$
$$8.1 \times 10^{-4} \text{ or } 0.81 \text{ permil,}$$

where the value 3.1 is the ^{238}U/^{235}U ratio in the solar system material 4.6×10^9 years ago.

The change in the abundance of ^{238}U can be similarly calculated to be 0.17 permil. The abundance of ^{235}U in the Allende inclusions is therefore expected to be enhanced relative to ^{238}U by 1.00081/1.00017 = 1.00064 or 0.64 permil. It appears that the results obtained by Chen and Wasserburg [7.90] are in good agreement with the theoretically predicted value.

Appendices

Appendix I. Goldschmidt's table of the abundance of the elements (originally compiled in 1938 and up-dated to 1954)

Z	Element	Lithosphere p.p.m.	atoms per 100 Si	Meteorites p.p.m.	atoms per 100 Si	Solar atmosphere[a] atoms per 100 Si	Stellar atmosphere[b] atoms per 100 Si
1	H	–	–	–	–	1 550.00	–
2	He	–	–	–	–		–
3	Li	65	0.1	4	0.01	0.0003	–
4	Be	6[c]	0.0067	1	0.0020	0.0002	–
5	B	10	0.0095	1.5	0.0024	0.32[d]	–
6	C	320	0.27	300	0.33	1000	–
7	N					2 100	–
8	O	466 000	296	323 000	347	2 800	–
9	F	800	0.43	3.3	0.03	0.32[d]	–
10	Ne	–					–
11	Na	28 300	12.4	5 950	4.42	98	100
12	Mg	20 900	8.76	123 000	87.24	170	71
13	Al	81 300	30.5	13 800	8.79	11	65
14	Si	277 200	100	163 000	100	100	–
15	P	1 200	0.391	1 050	0.58	0.03[d]	–
16	S	520	0.16	21 200	11.4	43	–
17	Cl	480	0.14	1 000–1 500[e]	0.4–0.6[e]		–
18	A	–					–
19	K	25 900	6.70	1 540	0.69	0.81	2.2
20	Ca	36 300	9.17	13 300	5.71	8.7	50
21	Sc	5[c]	0.0011	4	0.0015	0.011	–
22	Ti	4 400	0.92	1 320	0.47	0.47	1.3
23	V	150	0.030	39	0.013	0.058	0.3
24	Cr	200	0.039	3 430	1.13	1.95	0.8
25	Mn	1 000	0.18	2 080	0.66	1.48	2.1
26	Fe	50 000	9.13	288 000	89.1	270	12
27	Co	40	0.0069	1 200	0.35	0.55	–
28	Ni	100	0.0175	15 680	4.60	4.6	–
29	Cu	70	0.010	170	0.046	0.090	–
30	Zn	80	0.0124	138	0.0362	0.31	0.69

Z	Element	Lithosphere p.p.m.	atoms per 100 Si	Meteorites p.p.m.	atoms per 100 Si	Solar atmosphere[a] atoms per 100 Si	Stellar atmosphere[b] atoms per 100 Si
31	Ga	15	0.0022	35.6	0.088	0.0003[d]	—
32	Ge	7	0.000 97	55	0.0130	0.003	—
33	As	5	0.000 67	detected	—	0.000 019	—
34	Se	0.09	0.000 012	7	0.0015	—	—
35	Br	2.5	0.000 32	20	0.0043	—	—
36	Kr	—	—	—	—	—	—
37	Rb	280	0.033	0.45	0.0007	0.0002	—
38	Sr	150	0.017	20	0.004	0.011	0.035
39	Y	28.1[c]	0.003 07	4.72	0.000 974	0.008	—
40	Zr	220	0.026	73	0.0139	0.0012	—
41	Nb	20	0.002	—	—	0.000 03	—
42	Mo	2.3	0.000 24	5.3	0.000 95	0.0003	—
43	Ma	—	—	—	—	—	—
44	Ru	—	—	2.23	0.000 36	0.000 16	—
45	Rh	0.001	0.000 000 1	0.80	0.000 13	0.000 01	—
46	Pd	0.010	0.000 000 9	1.54	0.000 25	0.000 04	—
47	Ag	0.02	0.000 001 8	2	0.000 32	0.000 03	—
48	Cd	0.18	0.000 016	2.4	—	0.0005[d]	—
49	In	0.1	0.000 007 1	0.15	0.000 023	0.000 003[e]	—
50	Sn	40	0.003 43	20	0.0029	0.000 02	—
51	Sb	(1)	0.000 083	detected	—	0.000 02[d]	—
52	Te	(0.0018)[e]	—	(0.1)[e]	—	—	—
53	I	0.3	0.000 024	1	0.000 136	—	—
54	Xe	—	—	—	—	—[e]	—
55	Cs	3.2	0.0006	0.01	0.000 01	—[e]	—
56	Ba	430	0.0312	6.9	0.000 83	0.000 46	0.0082
57	La	18.3[c]	0.001 28	1.58	0.000 208	0.002	—
58	Ce	41.6[c]	0.003 21	1.77[e]	0.000 232[e]	0.0008	—
59	Nr	5.53[c]	0.000 389	0.75	0.000 0964	0.000 01[d]	—
60	Nd	23.9[c]	0.001 62	2.59	0.000 331	—	—

Appendix I (continued).

Z	Element	Lithosphere p.p.m.	atoms per 100 Si	Meteorites p.p.m.	atoms per 100 Si	Solar atmosphere[a] atoms per 100 Si	Stellar atmosphere[b] atoms per 100 Si
61	—	—	—	—	—	—	—
62	Sm	6.47[c]	0.000 419	0.95	0.000 115	0.0001	—
63	Eu	1.06[c]	0.000 068	0.25	0.000 028	0.000 08[d]	—
64	Gd	6.36[c]	0.000 394	1.42	0.000 165	0.000 04[d]	—
65	Tb	0.91[c]	0.000 056	0.45	0.000 052	—	—
66	Dy	4.47[c]	0.000 269	1.80	0.000 203	0.0001[d]	—
67	Ho	1.15[c]	0.000 068	0.51	0.000 057	—	—
68	Er	2.47[c]	0.000 144	1.48	0.000 163	0.000 004[d]	—
69	Tm	0.20[c]	0.000 011 5	0.26	0.000 029	0.000 01[d]	—
70	Yb	2.66[c]	0.000 149	1.42	0.000 150	0.000 03[d]	—
71	Lu	0.75[c]	0.000 037	0.46	0.000 048	0.000 03[d]	—
72	Hf	4.5	0.000 30	1.6	0.000 15	0.000 008[d]	—
73	Ta	2.1	0.000 117	—	—	0.000 003[d]	—
74	W	1	0.000 055	15	0.00145	0.000 005[d]	—
75	Re	0.001	0.000 000 054	0.0020	0.000 000 18	—	—
76	Os	—	—	1.92	0.000 174	0.000 01[d]	—
77	Ir	0.001	0.000 000 05	0.65	0.000 058	0.000 002[e]	—
78	Pt	0.005	0.000 000 27	3.25	0.000 287	0.0001	—
79	Au	0.001	0.000 000 05	0.7	0.000 057	—	—
80	Hg	0.5[c]	0.000 025	—	—	—	—
81	Tl	0.3	0.000 015	detected	0.000 015	—	—
82	Pb	16	0.000 80	11	0.00091	0.0018	—
83	Bi	0.2	0.000 009	detected	—	—	—
90	Th	11.5[c]	0.000 50	0.8	0.000 59	—	—
92	U	4	0.000 16	0.36	0.000 023	—	—

[a] After Unsöld (1948) and Russell, Dugan, and Stewart (1938)
[b] After Payne (1925)
[c] Indicates sediments
[d] Signifies doubtful value
[e] Very doubtful

150

Appendix II. The 1956 Suess-Urey abundance table for the individual nuclei

Element		A	N	I	Log H	H
1	H				10.60	4.00×10^{10}
		1	0	−1	10.60	4.00×10^{10}
		2	1	0	6.75	$5.7 \ \times 10^{6}$
2	He					
		3	1	−1	−	−
		4	2	0	9.49	3.08×10^{9}
3	Li				2.00	100
		6	3	0	0.87	7.4
		7	4	1	1.97	92.6
4	Be	9	5	1	1.30	20
5	B				1.38	24
		10	5	0	0.65	4.5
		11	6	1	1.29	19.5
6	C				6.56	3.54×10^{6}
		12	6	0	6.54	3.50×10^{6}
		13	7	1	4.59	3.92×10^{4}
7	N				6.82	6.60×10^{6}
		14	7	0	6.82	6.58×10^{6}
		15	8	1	4.38	2.41×10^{4}
8	O				7.33	2.14×10^{7}
		16	8	0	7.33	2.13×10^{7}
		17	9	1	3.90	8.00×10^{3}
		18	10	2	4.64	4.36×10^{4}
9	F	19	10	1	3.20	1600
10	Ne				6.93	$8.6 \ \times 10^{6}$
		20	10	0	6.89	7.74×10^{6}
		21	11	1	4.41	2.58×10^{4}
		22	12	2	5.92	8.36×10^{5}
11	Na	23	12	1	4.64	4.38×10^{4}
12	Mg				5.96	9.12×10^{5}
		24	12	0	5.86	7.21×10^{5}
		25	13	1	4.96	9.17×10^{4}
		26	14	2	5.00	1.00×10^{5}
13	Al	27	14	1	4.98	9.48×10^{4}
14	Si				6.00	1.00×10^{6}
		28	14	0	5.96	9.22×10^{5}
		29	15	1	4.67	4.70×10^{4}
		30	16	2	4.49	3.12×10^{4}
15	P	31	16	1	4.00	1.00×10^{4}
16	S				5.57	3.75×10^{5}
		32	16	0	5.55	3.56×10^{5}
		33	17	1	3.44	2.77×10^{3}
		34	18	2	4.19	1.57×10^{4}
		36	20	4	1.71	51
17	Cl				3.95	8850
		35	18	1	3.82	6670
		37	20	. 3	3.34	2180
18	A				5.18	1.50×10^{5}
		36	18	0	5.10	1.26×10^{5}
		38	20	2	4.38	$2.4 \ \times 10^{4}$
		40	22	4		

Element		A	N	I	Log H	H
19	K				3.50	3160
		39	20	1	3.47	2940
		40	21	2	0.58 − 1	0.38
		41	22	3	2.34	219
20	Ca				4.69	4.90×10^4
		40	20	0	4.68	4.75×10^4
		42	22	2	2.50	314
		43	23	3	1.80	64
		44	24	4	3.02	1040
		46	26	6	0.20	1.6
		48	28	8	1.94	87.7
21	Sc	45	24	3	0.43	2.8
22	Ti				3.39	2440
		46	24	2	2.29	194
		47	25	3	2.28	189
		48	26	4	3.25	1790
		49	27	5	2.13	134
		50	28	6	2.11	130
23	V				2.34	220
		50	27	4	0.74 − 1	0.55
		51	28	5	2.34	220
24	Cr				3.89	7800
		50	26	2	2.54	344
		52	28	4	3.81	6510
		53	29	5	2.87	744
		54	30	6	2.31	204
25	Mn	55	30	5	3.84	6850
26	Fe				5.78	6.00×10^5
		54	28	2	4.55	3.54×10^4
		56	30	4	5.77	5.49×10^5
		57	31	5	4.13	1.35×10^4
		58	32	6	3.30	1980
27	Co	59	32	5	3.25	1800
28	Ni				4.44	2.74×10^4
		58	30	2	4.27	1.86×10^4
		60	32	4	3.86	7170
		61	33	5	2.53	342
		62	34	6	3.00	1000
		64	36	8	2.50	318
29	Cu				2.33	212
		63	34	5	2.16	146
		65	36	7	1.82	66
30	Zn				2.69	486
		64	34	4	2.38	238
		66	36	6	2.13	134
		67	37	7	1.30	20.0
		68	38	8	1.96	90.9
		70	40	10	0.52	3.35
31	Ga				1.06	11.4
		69	38	7	0.84	6.86
		71	40	9	0.66	4.54

Element		A	N	I	Log H	H
32	Ge				1.70	50.5
		70	38	6	1.02	10.4
		72	40	8	1.14	13.8
		73	41	9	0.58	3.84
		74	42	10	1.27	18.65
		76	44	12	0.59	3.87
33	As	75	42	9	0.60	4.0
34	Se				1.83	67.6
		74	40	6	0.81	0.649
		76	42	8	0.80	6.16
		77	43	9	0.70	5.07
		78	44	10	1.20	16.0
		80	46	12	1.53	33.8
		82	48	14	0.78	5.98
35	Br				1.13	13.4
		79	44	9	0.83	6.78
		81	46	11	0.82	6.62
36	Kr				1.71	51.3
		78	42	6	0.24 − 1	0.175
		80	44	8	0.06	1.14
		82	46	10	0.77	5.90
		83	47	11	0.76	5.89
		84	48	12	1.47	29.3
		86	50	14	0.95	8.94
37	Rb				0.81	6.5
		85	48	11	0.67	4.73
		87	50	13	0.25	1.77
38	Sr				1.28	18.9
		84	46	8	0.03	0.106
		86	48	10	0.26	1.86
		87	49	11	0.12	1.33
		88	50	12	1.19	15.6
39	Y	89	50	11	0.95	8.9
40	Zr				1.74	54.5
		90	50	10	1.45	28.0
		91	51	11	0.79	6.12
		92	52	12	0.97	9.32
		94	54	14	0.98	9.48
		96	56	16	0.18	1.53
41	Nb	93	52	11	0.00	1.00
42	Mo				0.38	2.42
		92	50	8	0.56 − 1	0.364
		94	52	10	0.35 − 1	0.226
		95	53	11	0.58 − 1	0.382
		96	54	12	0.60 − 1	0.401
		97	55	13	0.37 − 1	0.232
		98	56	14	0.76 − 1	0.581
		100	58	16	0.37 − 1	0.234

Element		A	N	I	Log H	H
44	Ru				0.17	1.49
		96	52	8	0.93 − 2	0.0846
		98	54	10	0.52 − 2	0.0331
		99	55	11	0.28 − 1	0.191
		100	56	12	0.28 − 1	0.189
		101	57	13	0.40 − 1	0.253
		102	58	14	0.67 − 1	0.467
		104	60	16	0.43 − 1	0.272
45	Rh	103	58	15	0.33 − 1	0.214
46	Pd				0.83 − 1	0.675
		102	56	10	0.73 − 3	0.0054
		104	58	12	0.80 − 2	0.0628
		105	59	13	0.18 − 1	0.1536
		106	60	14	0.26 − 1	0.1839
		108	62	16	0.26 − 1	0.180
		110	64	18	0.96 − 2	0.0911
47	Ag				0.41 − 1	0.26
		107	60	13	0.13 − 1	0.134
		109	62	15	0.10 − 1	0.126
48	Cd				0.95 − 1	0.89
		106	58	10	0.04 − 2	0.0109
		108	60	12	0.90 − 3	0.0079
		110	62	14	0.04 − 1	0.111
		111	63	15	0.06 − 1	0.114
		112	64	16	0.33 − 1	0.212
		113	65	17	0.04 − 1	0.110
		114	66	18	0.41 − 1	0.256
		116	68	20	0.83 − 2	0.068
49	In				0.04 − 1	0.11
		113	64	15	0.66 − 3	0.0046
		115	66	17	0.02 − 1	0.105
50	Sn				0.12	1.33
		112	62	12	0.13 − 2	0.0134
		114	64	14	0.96 − 3	0.0090
		115	65	15	0.67 − 3	0.00465
		116	66	16	0.28 − 1	0.189
		117	67	17	0.01 − 1	0.102
		118	68	18	0.50 − 1	0.316
		119	69	19	0.06 − 1	0.115
		120	70	20	0.64 − 1	0.433
		122	72	22	0.80 − 2	0.063
		124	74	24	0.90 − 2	0.079
59	Sb				0.39 − 1	0.246
		121	70	19	0.15 − 1	0.141
		123	72	21	0.02 − 1	0.105

Appendix II (continued)

Element		A	N	I	Log H	H
52	Te				0.67	4.67
		120	68	16	0.62 − 3	0.00420
		122	70	18	0.06 − 1	0.115
		123	71	19	0.62 − 2	0.0416
		124	72	20	0.34 − 1	0.221
		125	73	21	0.52 − 1	0.328
		126	74	22	0.94 − 1	0.874
		128	76	24	0.17	1.48
		130	78	26	0.20	1.60
52	I	127	74	21	0.90 − 1	0.80
54	Xe				0.60	4.0
		124	70	16	0.58 − 3	0.00380
		126	72	18	0.55 − 3	0.00352
		128	74	20	0.88 − 2	0.0764
		129	75	21	0.02	1.050
		130	76	22	0.21 − 1	0.162
		131	77	23	0.93 − 1	0.850
		132	78	24	0.03	1.078
		134	80	26	0.62 − 1	0.420
		136	82	28	0.55 − 1	0.358
55	Cs	133	78	23	0.66 − 1	0.456
56	Ba				0.56	3.66
		130	74	18	0.57 − 3	0.00370
		132	76	20	0.55 − 3	0.00356
		134	78	22	0.95 − 2	0.0886
		135	79	23	0.38 − 1	0.241
		136	80	24	0.45 − 1	0.286
		137	81	25	0.62 − 1	0.414
		138	82	26	0.42	2.622
57	La				0.30	2.00
		138	81	24	0.25 − 3	0.0018
		139	82	25	0.30	2.00
58	Ce				0.35	2.26
		136	78	20	0.64 − 3	0.0044
		138	80	22	0.75 − 3	0.00566
		140	82	24	0.30	2.00
		142	84	26	0.40 − 1	0.250
59	Pr	141	82	23	0.60 − 1	0.40
60	Nd				0.16	1.44
		142	82	22	0.59 − 1	0.39
		143	83	23	0.24 − 1	0.175
		144	84	24	0.54 − 1	0.344
		145	85	25	0.08 − 1	0.119
		146	86	26	0.39 − 1	0.248
		148	88	28	0.91 − 2	0.0824
		150	90	30	0.91 − 2	0.0806
62	Sm				0.82 − 1	0.664
		144	82	20	0.32 − 2	0.0108
		147	85	23	0.00 − 1	0.100
		148	86	24	0.87 − 2	0.0748
		149	87	25	0.96 − 2	0.0920

Element		A	N	I	Log H	H
		150	88	26	0.69 − 2	0.0492
		152	90	28	0.25 − 1	0.176
		154	92	30	0.17 − 2	0.150
63	Eu				0.27 − 1	0.187
		151	88	25	0.95 − 2	0.0892
		153	90	27	0.99 − 2	0.0976
64	Gd				0.83 − 1	0.684
		152	88	24	0.14 − 3	0.00137
		154	90	26	0.17 − 2	0.0147
		155	91	27	0.00 − 1	0.101
		156	92	28	0.15 − 1	0.141
		157	93	29	0.03 − 1	0.107
		158	94	30	0.23 − 1	0.169
		160	96	32	0.17 − 1	0.149
65	Tb	159	94	29	0.98 − 2	0.0956
66	Dy				0.74 − 1	0.556
		156	90	24	0.46 − 4	0.00029
		158	92	26	0.70 − 4	0.000502
		160	94	28	0.10 − 2	0.0127
		161	95	29	0.02 − 1	0.105
		162	96	30	0.15 − 1	0.142
		163	97	31	0.14 − 1	0.139
		164	98	32	0.19 − 1	0.157
67	Ho	165	98	33	0.07 − 1	0.118
68	Er				0.50 − 1	0.316
		162	94	26	0.50 − 4	0.000316
		164	96	28	0.67 − 3	0.00474
		166	98	30	0.02 − 1	0.104
		167	99	31	0.88 − 2	0.770
		168	100	32	0.93 − 2	0.0850
		170	102	34	0.65 − 2	0.0228
69	Tm	169	100	31	0.50 − 2	0.0318
70	Yb				0.34 − 1	0.220
		168	98	28	0.48 − 4	0.00030
		170	100	30	0.82 − 3	0.00666
		171	101	31	0.50 − 2	0.0316
		172	102	32	0.68 − 2	0.0480
		173	103	33	0.55 − 2	0.0356
		174	104	34	0.84 − 2	0.0678
		176	106	36	0.44 − 2	0.0278
71	Lu				0.70 − 2	0.050
		175	104	33	0.69 − 2	0.0488
		176	105	34	0.11 − 3	0.0013
72	Hf				0.68 − 1	0.438
		174	102	30	0.90 − 4	0.00078
		176	104	32	0.35 − 2	0.0226
		177	105	33	0.91 − 2	0.0806
		178	106	34	0.07 − 1	0.119
		179	107	35	0.78 − 2	0.0604
		180	108	36	0.19 − 1	0.155
73	Ta	181	108	35	0.81 − 2	0.065

Element		A	N	I	Log H	H
74	W				0.69 − 1	0.49
		180	106	32	0.78 − 4	0.0006
		182	108	34	0.11 − 1	0.13
		183	109	35	0.84 − 2	0.070
		184	110	36	0.17 − 1	0.15
		186	112	38	0.14 − 1	0.14
75	Re				0.13 − 1	0.135
		185	110	35	0.70 − 2	0.0500
		187	112	37	0.93 − 2	0.0850
76	Os				0.00	1.00
		184	108	32	0.26 − 4	0.00018
		186	110	34	0.20 − 2	0.0159
		187	111	35	0.22 − 2	0.0164
		188	112	36	0.12 − 1	0.133
		189	113	37	0.21 − 1	0.161
		190	114	38	0.42 − 1	0.264
		192	116	40	0.61 − 1	0.410
77	Ir				0.91 − 1	0.821
		191	114	37	0.50 − 1	0.316
		193	116	39	0.70 − 1	0.505
78	Pt				0.21	1.625
		190	112	34	0.00 − 4	0.0001
		192	114	36	0.10 − 2	0.0127
		194	116	38	0.73 − 1	0.533
		195	117	39	0.74 − 1	0.548
		196	118	40	0.62 − 1	0.413
		198	120	42	0.07 − 1	0.117
79	Au	197	118	39	0.16 − 1	0.145
80	Hg				0.45 − 1	0.284
		196	116	36	0.65 − 4	0.00045
		198	118	38	0.45 − 2	0.0285
		199	119	39	0.68 − 2	0.0481
		200	120	40	0.82 − 2	0.0656
		201	121	41	0.57 − 2	0.0375
		202	122	42	0.93 − 2	0.0844
		204	124	44	0.29 − 2	0.0194
81	Tl				0.03 − 1	0.108
		203	122	41	0.50 − 2	0.0319
		205	124	43	0.88 − 2	0.0761
82	Pb				0.67 − 1	0.47
		204	122	40	0.80 − 3	0.0063
		206	124	42	0.09 − 1	0.122
		207	125	43	0.00 − 2	0.0995
		208	126	44	0.39 − 1	0.243
83	Bi	209	126	43	0.16 − 1	0.144
90	Th	232	142	52		
92	U	235	143	51		
		238	146	54		

Appendix III. The 1975 abundance table compiled by Virginia Trimble (Rev. Mod. Phys. 47. 877, 1975)

Z	Element	Meteorites	Solar photosphere	Q	Solar corona	Cosmic ray sources
1	H	3.18×10^{10}	2.5×10^{10}	a	2.5×10^{10}	4.1×10^{9}
2	He	2.21×10^{9}	2×10^{9}	b	2.0×10^{9}	3.1×10^{8}
3	Li	49.5	0.2	b	–	–
4	Be	0.81	0.2	b	–	–
5	B	3.2	<4.0	b	–	–
6	C	1.18×10^{7}	10^{7}	a	1.4×10^{7}	1.18×10^{7}
7	N	3.74×10^{6}	3×10^{6}	a	2.8×10^{6}	1.3×10^{6}
8	O	2.15×10^{7}	1.6×10^{7}	a	2.0×10^{7}	1.3×10^{7}
9	F	2450	1000	c	–	–
10	Ne	3.44×10^{6}	10^{6}	b	1.7×10^{6}	1.8×10^{6}
11	Na	6.0×10^{4}	5×10^{4}	a	5.3×10^{4}	9.4×10^{4}
12	Mg	1.061×10^{6}	8×10^{5}	a	9.4×10^{5}	2.7×10^{6}
13	Al	8.5×10^{4}	8×10^{4}	a	7.9×10^{4}	2.4×10^{5}
14	Si	1×10^{6}	10^{6}	a	1.1×10^{6}	2.4×10^{6}
15	P	9600	10^{4}	b	7100	2.4×10^{4}
16	S	5.0×10^{5}	4×10^{5}	a	3.5×10^{5}	3.5×10^{5}
17	Cl	5700	8×10^{3}	b	–	–
18	Ar	1.172×10^{5}	2.4×10^{4}	c	8×10^{4}	8.3×10^{4}
19	K	3790	8×10^{3}	c	1.4×10^{4}	–
20	Ca	6.25×10^{4}	6×10^{4}	a	6.3×10^{4}	2.6×10^{5}
21	Sc	35	30	b	315	–
22	Ti	2775	1600	b	5000	–
23	V	262	250	b	1.6×10^{4}	–
24	Cr	1.27×10^{4}	1.6×10^{4}	b	1.8×10^{4}	3.5×10^{4}
25	Mn	9300	6000	b	8900	2.4×10^{4}
26	Fe	8.9×10^{5}	6×10^{5}	b	8.2×10^{5}	2.6×10^{6}
27	Co	2210	800	b	5600	–
28	Ni	4.80×10^{4}	8×10^{4}	a	8.5×10^{4}	9.4×10^{4}
29	Cu	540	400	b	1000	7.4×10^{3}
30	Zn	1244	630	b	–	9750 (Z = 30–31)
31	Ga	48	16	b	–	
32	Ge	115	80	b	–	1.2×10^{3} (Z = 32–34)
33	As	6.6	no lines		–	–
34	Se	67.2	no lines		–	–
35	Br	13.5	no lines		–	113 (Z = 35–39)
36	Kr	46.8	no lines		–	
37	Rb	5.88	10	b	–	–
38	Sr	26.9	20	a	25	–
39	Y	4.8	1.6	b	–	–
40	Zr	15.1	16	b	–	61 (Z = 40–44)
41	Nb	1.4	5	b	–	–
42	Mo	4.0	8	b	–	–
44	Ru	1.9	2.5	b	–	–
45	Rh	0.4	0.8	b	–	0.9 (Z = 45–49)
46	Pd	1.3	0.6	b	–	–
47	Ag	0.45	0.2	b	–	–
48	Cd	1.48	2.5	b	–	–
49	In	0.189	1.3	b	–	–
50	Sn	3.6	0.8	b	–	14.4 (Z = 50–54)
51	Sb	0.316	0.25	b	–	–

Appendix III (continued)

Z	Element	Meteorites	Solar photosphere	Q	Solar corona	Cosmic ray sources
52	Te	6.42	no lines		–	–
53	I	1.09	no lines		–	–
54	Xe	5.38	no lines		–	–
55	Cs	0.387	<2.0	b	–	10.6 (Z = 55–59)
56	Ba	4.8	2.5	b	10	–
57	La	0.445	1.6	b	–	–
58	Ce	1.18	2.0	b	–	–
59	Pr	0.149	1.0	b	–	–
60	Nd	0.78	1.6	b	–	0.5 (Z = 60–64)
62	Sm	0.226	1.3	b	–	–
63	Eu	0.085	0.13	b	–	–
64	Gd	0.297	0.3	b	–	–
65	Tb	0.055	no f values		–	0.2 (Z = 65–69)
66	Dy	0.36	0.3	b	–	–
67	Ho	0.079	no f values		–	–
68	Er	0.225	0.16	c	–	–
69	Tm	0.034	0.06	b	–	–
70	Yb	0.126	0.16	c	–	1.2 (Z = 70–74)
71	Lu	0.036	0.16	b	–	–
72	Hf	0.15	0.2	c	–	–
73	Ta	0.021	no lines		–	–
74	W	0.16	10	c	–	–
75	Re	0.053	<0.01	b	–	13.4 (Z = 75–79)
76	Os	0.75	0.16	c	–	–
77	Ir	0.717	4.0	c	–	–
78	Pt	1.4	3.2	c	–	–
79	Au	0.202	0.13	b	–	–
80	Hg	0.4	<3	b	–	5.6 (Z = 80–84)
81	Tl	0.192	0.2	b	–	–
82	Pb	4	2.0	a	–	–
83	Bi	0.143	<2.0	b	–	–
90	Th	0.045	0.20	c	–	6.1 (Z = 90–94)
92	U	0.0262	<0.27	c	–	–

The meteoritic data are those of A.G.W. Cameron, Space Sci. Rev. 15:121 (1973); the photospheric data are those of O. Engvold and Ö. Hauge (Report No. 39, Inst. Theor. Astrophys. Oslo, 1974); the corona data are those of G.L. Withbroe (in: The Menzel Symposium on Solar Physics, Atomic Spectra, and Gaseous Nebulae, NBS Special Publication 353, edited by K.B. Gebbie, GPO, Washington, D.C. 1971, p. 127); the cosmic ray source data are those of M.M. Shapiro and R. Silberberg (in: Proceedings of the 12th International Conference on Cosmic Rays, Hobart, Tasmania, Vol. 1, 1971. p. 221) and P.B. Price (in: Cosmochemistry, edited by A.G.W. Cameron, Reidel, Dordrecht, 1973, p. 69). Q is the quality of the photospheric determinations as estimated by Engvold and Hauge (1974); a represents errors of less than 30%, b less than a factor of 2.5, and c less than a factor of 10

Subject Index

abundance
–, rare earth elements 10
–, Suess-Urey curve 71
–, of the elements
–, –, cosmic 11
–, –, Goldschmidt's table 148
–, –, Suess-Urey's table 151
–, –, Trimble's table 158
achondrite, Norton County 142
age of the elements 86, 88
Allende inclusion 128
Allende meteorite 117, 122–124
all-present-theory, of Noddack 15
antimatter, world of 80
α-process 71
argon 130
–, isotopic anomalies 130
Aston's whole number rule 8
atmophil elements 12
atomic weight 7

barium 140
–, isotopic anomalies 140, 141
–, isotopic anomalies 140, 141
Belgian Congo pitchblende 27, 36
beryllium
–, origin of 115
–, x-process 72, 115–117
bing-bang theory 64, 79, 80
biophil elements 12
black-body radiation, cosmic 78
boron
–, origin of 115
–, x-process 72, 115–117

calcium 145
cerium 144
chalcophil elements 12
Clarke's number 8
climatic effect, possible, supernova explosion
68

C-N cycle 62
cosmic abundance, of the elements 11
cosmic black-body radiation 78
cosmology, symmetric 80

d-d reaction 62
decay, interval
–, I/Xe 109, 111
–, Pu/Xe 109, 111
deficient elements 59
depleted uranium 37

earth, nuclear processes 33
element 43, discovery by artificial means 16
element 61, discovery by artificial means 16
elements
–, abundance
–, –, Goldschmidt's table 148
–, –, Suess-Urey's table 151
–, –, Trimbles table 158
–, age 86, 88
–, atmophil 12
–, argon 130
–, –, isotopic anomalies 130
–, barium 140
–, –, isotopic anomalies 140, 141
–, biophil 12
–, beryllium 115
–, –, x-process 72, 115–117
–, boron 115
–, –, x-process 72, 115–117
–, calcium 145
–, californium-254, supernovae 68
–, cerium 144
–, chalcophil 12
–, C-N cycle 62
–, cosmic abundance 11
–, deficient 59
–, discovery by artificial means
–, –, element 43 16
–, –, element 61 16

—, gadolinium 142
—, —, isotopic anomalies 142, 144
—, geochemical classification 12
—, helium
—, —, α-process 71
—, —, d-d reaction 62
—, —, in the sun 57
—, —, proton-proton-chain 62
—, illinium 16
—, iodine- 129
—, —, half-life 88
—, —, pitchblende 37
—, iodine isotopes in uranium solutions 37, 38
—, isotopic anomalies 144
—, I/Xe decay interval 109, 111
—, krypton 130
—, —, isotopic anomalies 130—133
—, lithium
—, —, origin 115
—, —, x-process 115—117
—, lithophil 12
—, masurium 11, 16
—, magnesium
—, isotopic anomalies 124, 128
—, —, isolation 24—26
—, —, isolation from pitchblende 24—25
—, —, long-lived isotopes 19
—, —, name 17
—, —, primordial 22
—, —, search for 23
—, molybdenium-99, uranium salts 26
—, neodymium 144
—, neon 129
—, —, α-process 71
—, —, isotopic anomalies 129
—, neon-E 119, 129
—, nuclear processes on the earth 33
—, oxygen
—, —, isotopic anomalies 123, 125
—, —, the puzzle of 125
—, palladium 145
—, promethium 15, 17
—, —, in pitchblende 29
—, —, in uranium salts 28
—, promethium-147 in Oklo reactor 49
—, plutonium-239
—, —, in nature 32
—, —, Oklo reactor 50
—, plutonium-244
—, —, in the early solar system 102
—, —, hypothesis 93
—, Pu/Xe decay interval 109, 111
—, rare earth, abundance 10
—, ruthenium isotopes in Oklo reactor 53
—, samarium, isotopic anomalies 128
—, siderophil 12
—, silver 145

—, strontium isotopes in pitchblende 35
—, superheavy 112
— synthesis in stars 70
—, technetium 15, 17
—, technetium-99
—, —, in Oklo reactor 53
—, —, in pitchblende 27
—, —, in stars 19
—, titanium 145
—, uranium 145
—, —, depleted 37
—, —, isotopic anomalies 145, 146
—, uranium-238 to -235 ratio
—, —, constancy 46
—, —, in nature 44, 45, 47
—, —, lunear samples 47
—, xenon 133
—, —, excess fission in Pasamonte meteorite 103
—, —, excess in meteorites 89
—, —, isotopic anomalies 133, 135, 138, 139
—, —, isotopic composition 10
—, —, isotopes in radioactive minerals 34
—, —, the puzzle of 138, 139
e-process 72
evolution of stars, theories on 67
excess fission xenon, in the Pasamonte
    meteorite 103
excess Xe, in meteorites 89
expanding universe 80
extinct radioactivity 87

fission
—, excess, xenon in Pasamonte meteorite 103
—, spontaneous, discovery 31, 32
frozen thermodynamic equilibria, concept of 58

gadolinium 142
—, isotopic anomalies 142, 144
geochemical classification of the elements 12
Great Bear Lake pitchblende 35, 36

Harkins, rule of 8, 10
helium
—, burning 71
—, α-process 71
—, d-d reaction 62
—, in the sun 57
—, proton-proton-chain 62
Hot Springs, Arkansas 33
hydrogen burning 71

I/Xe decay interval 109, 111
illinium 16

infinite multiplication constant 38, 40
isotope 8
isotopes
−, iodine in uranium solutions 37, 38
−, long-lived, technetium 19
−, *r*-process 72, 74−78
−, ruthenium, in Oklo reactor 53
−, *s*-process 72, 74−77
−, strontium, pitchblende 35
−, xenon, in radioactive minerals 34
isotopic anomalies
−, a unified theory 120
−, argon 130
−, barium 140, 141
−, gadolinium 142, 144
−, krypton 130−133
−, magnesium 124, 128
−, meteorites 117
−, neon 129
−, other elements 144
−, oxygen 123, 125
−, samarium 128
−, uranium 145, 146
−, xenon 133, 135, 138, 139
isotopic composition, xenon 10
iodine-129
−, half-life of 8
−, in pitchblende 37
iodine isotopes, in uranium solutions 37, 38

Johanngeorgenstadt pitchblende 40−42

krypton 130
−, isotopic anomalies 130−133

lithium
−, origin of 115
−, *x*-process 72, 115−117
lithophil elements 12
lunear samples, uranium-238 to -235 ratio 47

magic numbers 17
magnesium, isotopic anomalies 124, 128
masurium 11, 16
matter
−, antimatter, the world of 80
−, prote hyle 5
−, universal 5
metabolon 86
meteorite
−, Allende 117, 122−124
−, archondrite, Norton County 142−144
−, isotopic anomalies 117

−, Pasamonte 102−104
−, −, excess fission xenon 103
−, Richardton 90, 91, 97
−, Richardton 90, 91, 97
−, Santa Clara 145
−, siderophil elements 12
minerals
−, chalcophil elements 12
−, lithophil elements 12
−, pleochloric haloes 87
−, radioactive, xenon 34
−, terrestrial, technetium 20
−, uraninites 43
−, uranium-238 to -235 ratio in lunar samples 47
−, xenology 107
−, −, unsolved problems 107
molybdenium-99, uranium salts 26

natural reactors
−, Oklo
−, −, discovery of 48
−, −, operating conditions 52
−, −, plutonium-239 50
−, −, promethium-147 49
−, −, ruthenium isotopes 53
−, −, technetium-99 53
−, search for 54
−, Sudbury structure 55
−, theory of 40
neodymium 144
neon 129
−, α-process 71
−, isotopic anomalies 129
neon-E 119, 129
neutrinos, from the sun 81
neutron
−, resonance capture in pitchblende 39
−, *r*-process 47, 72, 74−78
−, *s*-process 72, 74−77
neutron star 79
Norton County
−, achondrite 142, 144
−, meteorite 143
nuclear fluid, overheated = ylem 64, 66
nuclear processes on the earth 33
nucleosynthesis, chronology of 100

Oddo-Harkins, rule of 9
Oklo reactor
−, discovery of 48
−, operating conditions 52
−, ruthenium isotopes 53
−, technetium-99 53

origin
–, lithium 115
–, beryllium 115
–, boron 115
oxygen
–, isotopic anomalies 123, 125
–, the puzzle of 125

palladium 145
particles
–, elementary, superheavy 114
Pasamonte meteorite 102–104
–, excess fission xenon 103
pitchblende
–, Belgian Congo 27, 36
–, Great Bear Lake 35, 36
–, iodine-129 37
–, Johanngeorgenstadt 40–42
–, promethium 29
–, resonance capture, neutrons 39
–, strontium isotopes 35
–, technetium 27
–, –, isolation 24, 25
–, plutonium-239
–, –, in nature 32
–, –, in Oklo reactor 50
pleochroic haloes 87
plutonium-244
–, hypothesis 93
–, in the early solar system 102
polyneutron hypothesis 65
p-process 72, 77
probability, resonance escape 39, 40
promethium 15, 17
promethium-147
–, in Oklo reactor 49
–, in pitchblende 29
–, in uranium salts 28
prote hyle 5
proton, p-process 72
proton-proton-chain 62
pulsar 79
Pu/Xe decay 109, 111

quarks in nature 114

radiation
–, cosmic black-body 78
radioactive minerals, xenon isotopes 34
radioactivity
–, extinct 87
–, Hot Springs, Arkansas 33
–, decay interval

–, –, I/Xe 109, 111
–, –, Pu/Xe 109, 111
rare earth elements, abundance of 10
reactor
–, Oklo
–, –, discovery of 48
–, –, operating conditions 52
–, –, plutonium-239 50
–, –, promethium-147 49
–, –, ruthenium isotopes 53
–, –, technetium-99 53
–, Sudbury structure 55
–, natural
–, –, search for 54
–, –, theory of 40
resonance capture of neutrons in pitchblende
   39
resonance escape, probability 39, 40
Richardton meteorite 90, 91, 97
r-process 72, 74–79
ruthenium isotopes in Oklo reactor 53

samarium, isotopic anomalies 128
Santa Clara meteorite 145
siderophil elements 12
silver 145
solar system, early, plutonium-244 102
spontaneous fission, discovery of 31, 32
s-process 72, 74–77
stars
–, C-N cycle 62
–, evolution, theories 67
–, helium burning 71
–, hydrogen burning 71
–, neutron 79
–, proton-proton-chain 62
–, pulsar 79
–, S-type 19
–, supernova explosion, possible climatic
   effect 68
–, supernovae, californium-254 68
–, synthesis of the elements 70
–, technetium 19
steady-state theory 79
strontium isotopes in pitchblende 35
S-type stars 19
Sudbury structure 55
Suess-Urey abundance curve 71
sun
–, helium 57
–, neutrinos 81
–, temperature 83
superheavy elementary particles in nature 114
superheavy elements in nature 112
supernova explosion, possible climatic effect 81

supernovae, californium-254 68
symmetric cosmology 80
synthesis of the elements in stars 70

technetium 15, 17
−, isolation 24−26
−, −, from pitchblende 24, 25
−, long-lived isotopes 19
−, in pitchblende 27
−, in stars 19
−, in terrestrial minerals 20
technetium-99 in Oklo reactor 53
temperature, high, e-process 72
temperature of the sun 83
thermal utilization factor 39, 40
thermodynamic equilibria, frozen, concepts of 58
thermonuclear reactions, rate of 60
titanium 145

universal matter 5
universe
−, big-bang theory 64, 79, 80
−, expanding 80
−, steady-state theory 79
uraninites 43

uranium 145
−, depleted 37
−, isotopic anomalies 145, 146
−, salts
−, −, molybdenium-99 26
−, −, promethium 28
−, solutions, iodine isotopes 37, 38
uranium-238 to -235 ratio
−, constancy of 46
−, in lunear samples 47
−, in nature 44, 45, 47

whole number rule, Aston 8

yelm 64, 66

xenology 107
−, unsolved problems 107
xenon 133
−, isotopic anomalies 133, 135, 138, 139
−, isotopic composition of 10
−, isotopes in radioactive minerals 34
−, the puzzle of 138, 139
x-process 72, 115−117

D. L. Kepert

Inorganic Stereo-chemistry

1982. 206 figures, 45 tables. XII, 227 pages
(Inorganic Chemistry Concepts, Volume 6)
ISBN 3-540-10716-9

Contents: Introduction. – Polyhedra. – Four-Coordinate Compounds. – Five-Coordinate Compounds Containing only Unidentate Ligands. – Five-Coordinate Compounds Containing Chelate Groups. – Six-Coordinate Compounds Containing only Unidentate Ligands. – Six-Coordinate Compounds [M(Bidentate)$_2$ (Unidentate)$_2$]. – Six-Coordinate Compounds [M(Bidentate)$_3$]. – Six-Coordinate Compounds Containing Tridentate Ligands. – Seven-Coordinate Compounds Containing only Unidentate Ligands. – Seven-Coordinate Compounds Containing Chelate Groups. – Eight-Coordinate Compounds Containing only Unidentate Ligands. – Eight-Coordinate Compounds Containing Chelate Groups. – Nine-Coordinate Compounds. – Ten-Coordinate Compounds. – Twelve-Coordinate Compounds. – References. – Subject Index.

The historical development of stereochemistry has been dominated by two molecular polyhedra, the tetrahedron and the octahedron. Compounds containing five, seven, eight, nine or ten groups attached to a central atom have much more complicated stereochemistries which can be described by a range of semiregular, non-uniform, and "chemical" polyhedra. These polyhedra may have different types of vertices, different edge lengths, and different sized faces which lead to a number of important chemical consequences. These include bond length and bond angle distortions, the stability of different isomers, and the ease of structural rearrangements in solution.

An important recent advance concerns the stereochemistry of molecules containing ring systems, which are extremely important throughout chemistry. Such molecules may not have stereochemistries corresponding to any of the usual polyhedra, but are intermediate between two different idealized polyhedra. The precise location of a particular molecule along this continuous range of stereochemistries depends upon the geometric design of the ring system, which includes the number of atoms in ring and the size of these atoms.

The simple techniques outlined in this work are the best way, and in most cases the only way, that such complicated structures with coordination numbers from four to twelve can be predicted.

Springer-Verlag
Berlin
Heidelberg
New York

A. F. Williams

A Theoretical Approach to Inorganic Chemistry

1979. 144 figures, 17 tables. XII, 316 pages
ISBN 3-540-09073-8

Contents: Quantum Mechanics and Atomic Theory. – Simple Molecular Orbital Theory. – Structural Applications of Molecular Orbital Theory. – Electronic Spectra and Magnetic Properties of Inorganic Compounds. – Alternative Methods and Concepts. – Mechanism and Reactivity. – Descriptive Chemistry. –Physical and Spectroscopic Methods. – Appendices. – Subject Index.

This book outlines the application of simple quantum mechanics to the study of inorganic chemistry, and shows its potential for systematizing and understanding the structure, physical properties, and reactivities of inorganic compounds. The considerable strides made in inorganic chemistry in recent years necessitate the establishment of a theoretical framework if the student is to acquire a sound knowledge of the subject. A wide range of topics is covered, and the reader is encouraged to look for further extensions of the theories discussed. The book emphasizes the importance of the critical application of theory and, although it is chiefly concerned with molecular orbital theory, other approaches are discussed. This text is intended for students in the latter half of their undergraduate studies. (235 references)

Springer-Verlag
Berlin
Heidelberg
New York